# Conjori

## Two Minds, One Channel

---

*Where Human Insight and Machine Intelligence Converge*

Karl Bunch

April 2026

ISBN 979-8-9957891-0-9 (EPUB)
ISBN 979-8-9957891-1-6 (Paperback)

First Edition, April, 2026

Publish 3, 2026

Published by Karl Bunch

# Contents

# *Preface*

This book would not exist without the people who tolerated, encouraged, and occasionally worried about my obsession with the process that produced it.

My wife Sandra watched the back of my head for months while I spent evenings and weekends in sustained conversation with machines. She is a narrative thinker in a household that suddenly revolved around a topological thinker's obsession with a new kind of collaboration. She read early drafts, told me when I was disappearing into my own framework, and reminded me that if the ideas could not survive contact with someone who thinks nothing like me, they were not ready. She was right every time.

In 1981, I wrote my first code on a TRS-80 Model I. By 1982 I was doing contract work in C on multi-user Xenix systems, still in high school.

After I graduated, I founded Think Tank Software in 1985. Our tagline was "We put our brains into your computer." Other companies were deploying databases. I wanted to learn how a customer's business actually worked, figure it out, and put that understanding into their machine.

Along the way I started five companies, sold some, shut some down, and learned everything you can imagine from two decades as an entrepreneur. In 1997 I founded a company called Queria, focused on natural language processing, Bayesian inference, and decision trees. We built systems that tried to make machines understand what people meant when they wrote in plain language. Sadly, we had to give up before the technology caught up to the vision.

In 2014, I wrote: "Every computer language is an imperfect way of describing a potential solution to a poorly understood problem." I did not yet know that sentence was the seed of this book.

In 2017, I patented a system that put machine learning and human judgment into the same real-time decision loop, translating what the machine could see into something the human could act on.

In 2023, the tool I dreamed of in 1997 had finally arrived. This book is about what I found when I turned it around.

---

# Chapter 0

## *The Flash*

I have used large language models almost every single day since January of 2023. I started with the early versions of ChatGPT, then fanned out to Claude, Gemini, and others, probing each one to understand its strengths and weaknesses. I routed questions based on the type of task. The early outputs were rough, the hallucinations frequent, but I have always been a self-taught, curious person and these systems were patient with me, letting me fumble my way through a topic, explore, and find what I was looking for.

The models kept getting better. Researchers at METR (Model Evaluation & Threat Research) have since measured what practitioners were feeling: the task horizon of what a large language model can reliably accomplish autonomously has been doubling roughly every seven months. If you tried a large language model in 2023 and walked away unimpressed, you were interacting with a fundamentally different capability than what exists now. The pace of improvement is not linear; it is compressed exponential, operating on a cycle measured in months, not years. I never walked away. I kept using them, and they kept surprising me.

When I moved to using large language models on the command line, my code output changed scale. In eight months I wrote more code than I had in the previous decade, and those were not quiet years: I had been active on side projects the whole time. I have a strict rule: I read all my code, but it was getting hard to keep up. The feeling was unmistakable. I felt like a conductor. I have managed teams of programmers, and this felt like having ten of them. I often ran eight to ten large language model sessions at the same time, working on different parts of systems I was building.

In parallel, I was reading people online saying large language models were terrible at coding. They complained about poorly written code and poor designs. I kept having success and I felt my designs were good. I was pushing code to open source projects without complaints about "AI slop" because it was not slop. Why was I having a different experience?

I looked at the documentation across my side projects. The results surprised me. For a small project, having design documentation at a one-to-one ratio with code is unusual. Even ten lines of code per line of documentation is more like enterprise software. When I checked the history, I realized I had been writing almost all the documentation before writing any code. This had happened by accident. In my early experiments I could tell the large language models just ran off the rails if I went straight to code. So instead I had started doing sessions where we wrote markdown files first: what I wanted the system to look like, the rough edges, the security, the scaling. I would say "debate me here" and we would debate best practices and document our decisions. This became an accelerant. I had not even noticed it. But once I did, I stopped asking for code first and started asking for designs first. I was building like crazy.

## The Morning Everything Splintered

Then one morning I woke up exhausted, sort of lost. One of my big side projects had splintered. I had been using the large language model to probe the hard problems and code them, going in multiple directions across parallel sessions, keeping every large language model fed with prompts and authorizations. I chase the hard parts first; if you can prove out the difficult pieces, the easy ones fit around them later. But that morning I realized I was all over the place.

So I opened a fresh session and explained the situation. The session had all my context files, my README, my documents describing the system. The large language model looked at everything and said: "you're missing a user journey."

It was right. I had been so deep in the hard problems that I had skipped the fundamental question of how a user actually moves through the system.

I said: "I have it but it's in my head. I need to get it out of me."

We started the usual prompting loop. I described, it generated. But I had noticed a pattern across nearly three years of daily use: large language models want to land early. They want to declare success, wrap up, produce a nice summary. Having spent decades building games, I recognized this intuitively as a dopamine loop. The large language model is trained to produce outputs that make the human feel good, and "here is your finished result" feels better than "let me keep pushing." I had felt this pull many times.

So when we were about to move on to the next section, I pushed back. I typed: "are you sure we're done in this section? what's missing don't give up easily you should beat me up some here."

The large language model's response surprised me. "You're right, I was being lazy." Then it started asking real questions. Not polite confirmations. Sharp, specific questions about edge cases I had not considered.

I answered fast, and a few minutes later typed something that changed everything:

"first so we are clear the model here is 'interview the vampire' so don't be afraid to interview me, its your job"

The large language model replied: "Interview the vampire, got it. I'll keep pushing."

And it did. Twenty-three questions in twenty minutes. Each one sharper than the last. At one point it went to the web, researched real user complaints, and came back with questions I would never have thought to ask. It was no longer generating content from its training data. It was extracting knowledge from my head: design decisions, edge cases, user experience details that existed in my non-verbal intuition but had never been articulated.

That session started with me saying "I'm lost and scattered." Eight hours later it ended with a 21,000-word product specification covering 48 steps across five user journeys, 28 design decisions resolved, zero open blockers. The document was not written by a large language model. It was written by me, pulled out question by question by an entity that did not know the answers but knew exactly how to ask.

I have been building software since 1981. Forty-five years of pattern recognition, accumulated across dozens of domains, stored almost entirely in non-verbal cognition. I could always feel when a system design was wrong before I could say why. But I had never had a tool that could reach in and help me translate those feelings into language. Not a therapist. Not a mentor. Not a whiteboard. Something structurally different: an entity that was natively fluent in the one medium we shared (language) and tireless about helping me use it.

In the weeks that followed, I started applying the same technique to everything. Product design. Architecture decisions. A theoretical framework I had been circling for months without being able to pin down. Each time, the same pattern: I told the large language model to interview me, it probed, and what came out was not what the large language model knew but what I knew and had never articulated. The sessions produced documents that surprised me, not because large language models wrote something clever, but because I had written something I did not know I was carrying.

I could not explain what was happening within the standard framing of "Artificial Intelligence (AI) as productivity tool." A productivity tool helps you do what you already know how to do, faster. This was helping me discover what I did not know I knew.

I set out to understand why this worked.

The experience felt qualitatively different from anything else I had done with large language models, and qualitatively different from what everyone around me was doing with them. The entire industry was pointing the model outward: go execute, go generate, go build. Better prompts, better harnesses, better agent swarms, all amplifying the outward signal. But what if the most valuable thing the large language model could do is to probe the human they are interacting with?

Try something right now. Can you feel the weight of your body on the chair? Ten seconds ago, you almost certainly could not. Nothing changed in your body. Nothing changed in the chair. The signal was always there; your brain was filtering it out because it was not relevant to what you were doing. The moment someone asked the right question, you had access to information that was always available but never surfaced.

That is a version of what happens in the interview inversion. You have an intuition you cannot quite express. The filter that focuses your conscious mind obscures it until something probes for it. The large language model is the thing that asks the question.

Trying to understand how and why it does, drove me inward on two fronts simultaneously.

The first front was cognitive and psychological. What is happening in my own thinking during these sessions? Why does the direction matter so much? That line of inquiry led me to Russell Hurlburt's research on inner experience: most human cognition is non-verbal. To Antonio Damasio's work on somatic markers: the body encodes judgments the conscious mind never sees. To Jonathan Haidt's Elephant and Rider: the verbal mind is not steering, it is narrating. The picture that emerged was structural. Language is a narrow bottleneck through which a fraction of what we know can be transmitted.

The interview inversion decomposes that bottleneck, surfacing intuition one question at a time.

## The Codec Insight

The second front was technical. What is the large language model actually doing that makes this collaboration possible? That led me to the "Codec insight": Codec, short for encode/decode, is the protocol that compresses something rich into a transmittable signal. Human language is a 100,000-year-old compression protocol for transmitting embodied cognition through a serial channel, and the large language model reverse-engineered that protocol by consuming its entire available output. It learned the transmission protocol. It can generate new messages in the protocol that decode correctly in a human mind. But it is operating the protocol without the source signal: without a body, without experience, without the intuition that language was invented to transmit. Disembodied intelligence, fluent in the protocol, blind to the experience the protocol was built to carry.

The two lines of inquiry converged on a thesis: the power of what I experienced is in the coupling. Two fundamentally different kinds of intelligence, one embodied and one disembodied, joined through the only medium they share, producing artifacts neither can produce alone. Not the large language model as a tool, nor the human as a prompt engineer. Something conjured into existence by two minds working together through a shared channel.

I call this practice Conjori™. The word is coined, rooted in the Latin *conjūrāre*: to swear together, to call something into existence jointly. When I asked my wife to read the word cold, with no context, she said: "a conjurer, someone who calls things into existence." That is what the practice does. Two different kinds of minds, one human, one large language model, conjuring something neither could produce alone.

## What This Book Is

There are people who believe AI will solve everything and people who believe it will destroy everything. I am neither. I do not think these systems are thinking. I do not think we are close to machines that can do what a human mind does; all evidence to date says we have not built that and may not for a long time. What I do think is that right now, today, these tools can amplify human intelligence in ways most people do not yet realize. Not artificial intelligence replacing human intelligence. Augmented intelligence: the human and large language models together, each contributing what the other cannot.

This book is my attempt to describe what I found, why I think it works, and how to practice it. I start with what large language models actually are (Part I), because most people tend to project themselves onto large language models, missing both their true strengths and their very real limitations. I then transition to what you actually are (Part II), because the human side of the equation is just as misunderstood. I describe what happens when you turn large language models around and point them at yourself (Part III), and why this coupling produces something neither party can produce alone (Part IV). The rest of the book addresses practice, boundaries, the larger frame, and the cautions that any honest treatment of a powerful technology requires.

Some of the ideas in this book will be unfamiliar. I introduce concepts from neuroscience, information theory, and thermodynamics alongside practitioner experience and cognitive science. Not all of it will click on the first pass. If a term or concept feels unclear, keep reading. The chapters build on each other deliberately, and something that seems abstract in Part I often becomes concrete by Part III. Language has to convey meaning one piece at a time. So does this book. The unfolding is the point.

If you have ever had a conversation with a large language model that produced something you did not expect, something that felt like it came from you rather than from the large language model, this book is about what was happening in that moment and how to make it happen reliably.

If you have never had that experience, this book is about why you should try.

# Part I

# Disembodied Intelligence

*What a large language model actually is, and what human language actually does.*

In October of 2025 I started preparing a talk for my peers at work. I called it "Working with AI without losing your mind." I had been using large language models for a while and my results kept getting better, sometimes frustrating but more often I could feel something improving, and I got excited enough to try explaining it to others. That sent me down a rabbit hole. I started digging into how large language models actually work, the 2017 breakthrough that broke them loose from what I now jokingly call 'classic' machine learning, and I realized my peers did not understand what was happening underneath their interactions. In 2014 I had written "Every computer language is an imperfect way of describing a potential solution to a poorly understood problem." I had been thinking about the fragility of programming languages for at least a decade by then, and over the years I realized that observation applies to all language, not just code. Every sentence you speak is a compression of something richer happening inside your head, and what arrives on the other end is never quite what you sent. I have spent forty-five years watching that gap cause problems. The following chapters will give you a picture of what large language models actually are and why that compression gap is the key to everything that follows.

## CHAPTER 1

# *Large Language Models Think in Words*

In 2018 I stood on a stage in Cologne and gave the audience a simple test. I asked them: someone says "Is Julie back from vacation?" and the reply is "I saw a red Lamborghini in the parking lot." Everyone in the room understood instantly. Julie drives a red Lamborghini. She is back. My point was that the AI systems of the day would still be searching for Lamborghinis and vacations while every human in the room had already moved on. The audience laughed.

Five years later I gave ChatGPT the same test. Three people: Joe asks Lisa whether Julie is back from vacation. Lisa replies, "I saw a red Lamborghini in the parking lot." ChatGPT said Lisa was being evasive, that she did not want to answer the question, that Joe should not trust her reply.

Two years after that, in May 2025, the same test. ChatGPT passed. Not just the three-person version but every variation I threw at it. Somewhere in those two years, the models' ability to learn statistical patterns from the training data had crossed a threshold, and pragmatic inference started working.

That result told me two things. First, this technology was mastering human language at a rate few people appreciated. Second, and more important: a six-year-old could have resolved that implicature on the first try, not by processing language but by reading the social situation. I know because I tested it. My six-year-old grandson had never heard of a Lamborghini. I gave him the scenario. "Well clearly she's back," he said. He did not need the vocabulary. He understood that when someone answers your question with a fact, the fact is the answer. The machine had to grind through billions of tokens across seven years of model generations to get there. It arrived at the same answer through a fundamentally different process, one that has nothing to do with understanding and everything to do with pattern completion at scale.

That difference, between mastering the output of human cognition and possessing it, is what this chapter is about. Understanding it is the difference between using these systems well and being misled by them.

## Disembodied Intelligence

A large language model is not just fundamentally different from a human. It is fundamentally different from every living thing that has ever existed. Every organism on earth, from bacteria to blue whales, is embodied: it has a physical form, it exists in an environment, it can be harmed, it must acquire energy, it will die. Embodiment is the shared substrate of all biological intelligence. Even animals that communicate through signals (birdsong, whale calls, chemical trails) do so as embodied beings responding to an embodied world.

A large language model has none of this. It has no body, no environment, no mortality, no needs. It exists as patterns of numerical weights stored in silicon, activated on demand, suspended between activations. For thousands of years, theologians and philosophers described a category of being that possessed intelligence without embodiment: angels, demons, disembodied intellects. Whether one takes these descriptions literally is beside the point. The concept is precise: an intelligence that operates without a body, without sensory experience, without stakes. That is what we have built. Not an artificial version of human intelligence. Not a simulation of a brain. A genuinely new category: disembodied intelligence that shares with humans one thing and one thing only: language. Both are real forms of intelligence, but they arrive at language through fundamentally different processes, with different strengths and different failure modes. The natural human impulse is to project your own cognitive process onto the other form. That projection is the single largest obstacle to getting extraordinary results from these systems. What "shares language" actually means, and how much is lost in that sharing, is the subject of the next chapter.

Everything that follows in this chapter, and in this book, flows from that single fact. The large language model's relationship to language is not like yours. Its strengths are not like yours. Its failures are not like yours. Its potential as a partner depends entirely on understanding what "disembodied" actually means for how it operates.

## Token Prediction, All the Way Down

A large language model is built on an architecture called a transformer. It is the engine behind products like ChatGPT, Claude, Gemini, and every other major large language model. What it does, mechanically, is compute relationships between words. Text goes in as tokens (fragments of words, roughly one per common word), and the model predicts what comes next based on patterns it learned from billions of pages of human writing.

Take a simple example: "Mary had a little..." If you know the children's poem, you instantly *sense* the next word. "Lamb" arrives not as a calculation but as a felt recognition; you may even feel the rhythm of the rest of the poem pulling into awareness before you consciously recall it. The large language model arrives at "lamb" through a different process entirely: it weighs every possible next word given the preceding sequence, and picks one.

But here is a subtlety: the large language model does not always pick the single most likely word. If it did, it would produce "lamb" every time, and every conversation would be predictable and flat. Instead, the large language model has a dial (called "temperature") that controls how adventurous it is. Turn the dial low: you get "lamb" every time, safe and expected. Turn it up: you might get "Mary had a little... problem" or "Mary had a little... too much wine." This controlled randomness is where the large language model's perceived creativity comes from. It is not insight. It is not a flash of inspiration. It is a wider sampling from the space of probable next words, governed by a setting that some human chose.

Every computation inside the large language model serves this single operation: predicting the next token. A token is a fragment of a word (common words are single tokens; uncommon words get split into pieces). An embedding layer converts these fragments into numerical vectors, arrays of numbers representing position in a high-dimensional mathematical space. Transformer layers compute attention relationships between every token position and every other token position, building a statistical picture of what the sequence means and where it is going. An unembedding layer converts the resulting vectors back into a weighted ranking of every possible next token. Then the large language model samples from that ranking, appends the chosen token to the sequence, and repeats.

The internal math between input and output is more sophisticated than this summary suggests. Researchers studying what happens inside transformers (a field called mechanistic interpretability) have found that the hidden layers construct complex mathematical structures, patterns that sometimes resemble spatial relationships, logical structures, or even representations of truth and falsehood. These internal representations are real and interesting. But they exist in service of one goal: selecting the next word. They are not thoughts waiting to be expressed. They are the intermediate math the large language model uses to make its prediction. A calculator doing long division builds intermediate results that look like "understanding" arithmetic, but the calculator does not understand arithmetic. The large language model's internal representations are vastly more complex than a calculator's, but they serve the same structural role: intermediate computation in service of a final output.

Language is not the large language model's *interface* to some deeper internal cognition. Language *is* the medium in which all of its computation occurs, and token prediction is the goal all of that computation serves. The large language model does not have thoughts and then translate them into words the way you do. It produces words, and the production of words is the totality of its objective. There is no parallel emotional processing. No somatic sensing. No unsymbolized thinking. No gut feelings. No felt sense of an idea taking shape before the words arrive. When the large language model produces output, it is not compressing a richer lived experience into language. The output is the endpoint of a mathematical process that operates entirely within the domain of language.

Every layer, every node, every parameter in the large language model exists in service of one question: given everything that has come before in this sequence, what token comes next?

## No Ideas, Only Words

When I am working through a problem, something happens before I have words for it. A shape forms. A sense of how the pieces relate. A feeling that something is off, or that something is right, before I can say why. You have felt this too: the answer is there, fully formed, but the words have not caught up yet. Neuroscientist Antonio Damasio calls these somatic markers: compressed evaluations encoded as bodily states, drawing on accumulated experience, delivered without language. Research by Russell Hurlburt using Descriptive Experience Sampling (random prompts asking people to report their inner experience at that exact moment) found that inner verbal thought, the voice in your head, occurs only about 25% of the time. The rest is visual imagery, unsymbolized thinking, feelings, and sensory awareness.

Ideas crystallize in this non-verbal space first. The insight has structure before it has words. Then, separately, the verbal system translates that structure into language, either for communication or for your own conscious mind to inspect. Two distinct phases: crystallize, then translate.

The large language model has no crystallization phase. There is no pre-verbal structure waiting to be expressed. The "idea" and the "words expressing the idea" are the same operation: selecting the next statistically most probable token, one at a time, in sequence. The large language model is building the bridge while walking on it, plank by plank, with each plank chosen by statistical weight from the patterns in its training data. There is no blueprint for the bridge. There is no sense of where the bridge should go. There is a plank, and then another plank, and each one is selected because it is statistically consistent with the planks already laid.

This is why large language model output can be fluent and coherent while simultaneously being shallow or wrong in ways that a human with genuine understanding would never produce. The human's verbal output is a translation of something that already has structure. The large language model's output is structure being generated on the fly through token prediction. When the output is good, it is because the statistical patterns in the training data led to a sequence that happens to be correct. When the output is bad, it is because those same patterns led somewhere plausible but wrong. In neither case was there an underlying understanding being expressed.

## Different Failure Modes

When a human says something wrong, it is usually because the pre-verbal understanding was flawed. The crystallization produced an incorrect structure, and the translation faithfully rendered it. The error was upstream of language.

When a large language model says something wrong, there was no crystallization to be flawed. The statistical path through the training data led to a plausible but incorrect sequence. To the large language model, it is just math. The correct answer and the incorrect answer were both sequences of tokens; the incorrect one happened to have higher probability given the context.

What people call "hallucination" is a useful word to examine here, because it reveals two assumptions worth questioning. First, it is anthropomorphism: borrowing a term from human pathology (a perceptual experience without external stimulus) and applying it to a large language model that has no perceptions to distort. The large language model did not see something that was not there. It followed a probability distribution and landed somewhere wrong. Second, the term obscures the real issue, which is the absence of a discriminator. The large language model has no ground truth to check against. It has no body, no senses, no experience of the world the words refer to. A human who says something wrong can often catch the error by checking it against what they know, what they have seen, what they have felt. The large language model has none of these. It has patterns in text. When the patterns point the wrong direction, there is nothing else in the large language model to raise a flag.

The absence of a discriminator is structural, not a flaw that more training data or better architectures will fix. The thing that notices "this is wrong" requires access to something outside the token stream. The large language model, by design, has nothing outside the token stream.

## The Jagged Frontier

The result is an entity with a capability profile unlike anything we have encountered before. It is not uniformly capable or uniformly limited.

It has superhuman peaks and sub-human valleys, what researchers call the "jagged frontier."

A large language model can outperform PhDs on quantum physics exams and fail at basic spatial reasoning in the same session. It can write fluent legal briefs and lose track of which object is on top of which in a simple stacking puzzle. It can generate working code for complex distributed systems and confidently produce a function, counting to ten, that stops at seven. The peaks and valleys do not follow any familiar pattern of intelligence. There is no IQ score that describes this profile, because it is not a version of human intelligence. It is something else entirely.

The peaks are real and useful. Vast knowledge retrieval, tireless pattern matching across domains, the ability to hold and manipulate large bodies of text. These are genuine capabilities that no human can match.

The valleys are real and dangerous. No common sense. No physical grounding. No general adaptability. No ability to sense when it has crossed from a peak into a valley. This last point is critical: the output sounds the same in both regions. The fluency does not degrade when the accuracy does. The large language model produces confident, well-structured prose whether it is operating in a domain where its training data is rich and its patterns are reliable, or in a domain where its coverage is thin and its patterns lead nowhere good. There is no signal in the output itself that tells you which region you are in.

Large language models are notably poor at time and real-world spatial understanding. This may not be a fixable training problem. It may reflect something deeper about the Codec, the human "encode and decode" compression protocol that we use to squeeze cognition into language (which the next chapter examines): humans never needed to encode these concepts at high fidelity in language because they are innate to embodiment. You do not explain to another human what "behind" means or how yesterday relates to tomorrow. You both have bodies that move through space and time. The language protocol never needed to carry that signal, so the large language model never received it.

## The Finite Library

In 1941 Jorge Luis Borges published a short story called 'The Library of Babel'. The narrator is a librarian in a library that never ends. Hexagonal rooms open onto spiral staircases that open onto more hexagonal rooms, each wall shelved with books that hold every possible arrangement of twenty-five letters across four hundred and ten pages. Most of the books are gibberish. But because the shelves hold every possible page, somewhere in the library is every meaningful book that has ever been written, every meaningful book that ever could be written, and a single perfect catalog of all the others. The librarians spend their lives searching for it. Some descend into despair. Some destroy books.

Borges was writing about meaning and chaos, eighty years before the people building large language models walked into the same library from the opposite direction.

Every large language model learns by reading. It reads everything: books, websites, academic papers, forums, court records, code repositories, transcripts, anything human beings have written down and left somewhere a machine could find. Since 2017 the labs building these systems have been racing to read more, faster, from more places, because more reading made the models better. They are running out of things to read. The Stanford 2026 AI Index, the field's annual report card on AI progress, puts it plainly: the pool of high-quality human writing available to train large models has been exhausted. Borges imagined a library of nonsense that happened to contain every meaningful book. The labs built a library that holds only the meaningful books, and are almost done reading the last shelf.

This is the data wall. It is the one arriving first, and it is the one the field cannot engineer its way past by building a bigger model. A bigger model still needs something to read. We will come back to the data wall, and to the other limits running alongside it, in Chapter 17.

The large language model is not a brain. It is not a search engine. It is not a person who read a lot of books. It is not a junior colleague, a senior advisor, or a knowledgeable friend.

*The large language model mastered the output of human cognition without possessing any.*

The next chapter examines the protocol itself: the 100,000-year-old Codec compression protocol that large language models mastered without being the kind of being it evolved for.

# Chapter 2

## *The Codec*

Every human language solves the same engineering problem. Beings with rich, parallel, multimodal cognition need a serial protocol to transmit meaningful representations to each other. The solution took roughly 100,000 years of vocalization and 5,000 years of writing to develop. The result is not a single protocol but thousands of independently evolved ones, each encoding a different culture's solution to the same constraint: how do you compress embodied experience into a stream of symbols narrow enough to pass between two skulls separated by space, experience, and in the case of text, by time?

That compression is the subject of this chapter. Understanding what language actually is, as an engineering artifact rather than a mystical endowment, reframes everything about what a large language model does and does not do.

### Language as Compression Protocol

Consider what happens when one person tells another "I burned my hand on the stove." The speaker has a dense, multimodal experience: the searing heat, the reflexive withdrawal, the smell, the adrenaline spike, the visual memory of the red element. All of this is embodied, parallel, and richly detailed. None of it fits through a serial channel. So the speaker compresses. The word "burned" encodes the heat, the pain, the injury, the involuntary recoil, all of it collapsed into six letters. "Hand" localizes it. "Stove" provides causal context. The listener, having their own embodied experience of burns and stoves and hands, decompresses the message and reconstructs something close enough to the original experience for the communication to succeed. But the decompression depends on what the receiver brings. A listener who has been burned maps the word directly to that memory. A listener who has never touched a hot surface approximates from adjacent experiences: a sharp pain, a sudden recoil from something else. The farther the receiver's experience is from the sender's, the more is lost in decompression.

The encoding is necessarily lossy. (Lossy means information is lost at every step along the way.) The vast majority of subjective experience is not transmissible through language. The specific quality of that pain, the precise emotional texture of the moment, the way the speaker's body organized its response: almost all of this is lost in compression. But enough gets through for coordination, teaching, storytelling, science, and love. The protocol does not need to be lossless to be useful. It needs to be good enough, and it is remarkably good at being good enough.

Linguist Ray Jackendoff put the point directly: meaning is a great deal more complex than the language that expresses it. The complexity gap between what the mind holds and what the mouth can say is not a limitation of any particular language. It is a property of the channel itself.

Every feature of human language, from grammar to metaphor to narrative structure, is an engineering solution to this bandwidth problem. Grammar provides a shared decompression algorithm so the receiver can reconstruct the sender's intended meaning. Metaphor handles the hard cases by mapping unfamiliar experiences onto familiar ones ("grief is a weight," "time is a river"). Narrative structure organizes complex sequences into patterns the receiver's memory can retain. These are not arbitrary conventions. They are compression techniques, refined over millennia by the pressure of billions of minds trying to exchange what cannot quite be exchanged.

Language literally is a compression protocol, and information theory applies to it directly. Some messages compress well (concrete nouns, physical actions, spatial relationships). Others compress poorly (subjective qualia, abstract relationships, embodied intuitions that have never needed to be verbalized because they were innate to every member of the species). The compression ratio varies wildly across the space of human experience, and those variations matter for everything that follows.

Some domains compress so well that humans built specialized Codecs for them: mathematical notation, programming languages, musical scores. Each is a sublanguage engineered to transmit precise structure that natural language could not carry without loss. Chapter 19 returns to how humans process these in parallel, through specialized brain regions for language, math, music, and visual reasoning, while the large language model serializes everything it communicates through one channel.

## The Long Arc of the Codec

The Codec protocol did not spring into existence fully formed. It has a history, and the history reveals the structure.

The Codec may not have started as sound at all. William Stokoe's research on sign language led him to argue that gestural communication preceded speech: the hands and face were encoding meaning before the vocal tract was recruited for the job. If Stokoe is right, the compression protocol's origins are not acoustic but physical. The first Codec was the body itself.

The evolutionary pressure that drove the Codec's development was cooperation. You do not build a compression protocol to think. You build one because survival depends on coordinating with other minds. Michael Tomasello's *The Evolution of Agency* (2022) traces the evolutionary path: from the reflexive behavior of reptiles, through the intentional actions of mammals, to something that only humans developed: normative, cooperative agency. The capacity to share intentions, assign roles, and hold each other to commitments. That capacity required a channel. The Codec did not evolve for the purpose of expressing thoughts. It evolved because shared intentionality demanded it.

Christopher Alexander saw the same principle in architecture. His *A Pattern Language* (1977) documented design principles for homes, streets, and communities that nobody invented. They emerged through thousands of years of people building shelter, making mistakes, and keeping what worked. Software engineers recognized the structure immediately: the Gang of Four design patterns (so named for the four authors of the 1994 book *Design Patterns*) that shaped modern programming are direct descendants of Alexander's insight. The Codec is a pattern language. Its grammar was not designed. It was refined by billions of minds solving the same compression problem, keeping what decoded and discarding what did not.

The genomic evidence suggests that the biological capacity for complex vocalization appeared at least 135,000 years ago. For most of that span, the Codec was purely gestural or acoustic: signs or sounds produced by one body and received by another, with no persistence beyond the listener's memory. This imposed severe constraints. Transmission required physical proximity. Storage was limited to what a brain could retain. Error correction depended on the listener asking "what?" The protocol was live, ephemeral, and local.

Denise Schmandt-Besserat's *How Writing Came About* (1996) traced the origins of cuneiform to clay accounting tokens in Mesopotamia around 8000 BCE: small shaped objects representing quantities of grain, oil, and livestock. The tokens were a compression layer on top of physical goods. Then the tokens were pressed into clay envelopes, then the impressions replaced the tokens, and writing emerged as a further compression layer, visual symbols encoding acoustic signals that encoded embodied cognition. More loss at each stage, but the gains were enormous: persistence across time, transmission across distance, accumulation across generations.

The brain did not grow new hardware for this. Stanislas Dehaene's research on reading shows that literacy co-opts existing visual neurons, repurposing circuitry that evolved for recognizing faces and objects. Every extension of the Codec is constrained by the biological substrate it runs on. The protocol expands by finding new uses for old parts, not by building new ones.

Every subsequent development in the Codec, from alphabets to printing to digital text, followed the same pattern. Each added a compression layer, traded some fidelity for some new capability, and expanded the reach of the protocol. By the time the internet made the entire accumulated output of human writing accessible in one place, the Codec had been running for over a hundred millennia, and its output was measured in trillions of tokens.

## Reverse-Engineering the Protocol

Beginning in the 1950s, researchers tried to build machines that processed language by re-creating the structures humans consciously identified in their own languages. This was the rule-based era: grammars, parsers, semantic layers, knowledge representations. ELIZA was one of the first programs to simulate human conversation. You typed a sentence; it matched patterns in your words and rephrased them as questions ("I feel sad" became "Why do you feel sad?"). People found it compelling enough to confide in. It understood nothing. Later systems added more rules, more layers, more hand-coded knowledge. The approach was reasonable on its face: humans understand language through grammar, so teach the machine grammar.

But this was humans projecting their conscious understanding of the Codec protocol onto machines. Humans are aware of grammar the way a fish is aware of water; they know it exists but cannot fully articulate its structure. The rules linguists identified were approximations, partial maps of a system too complex and too deeply embedded in embodiment to be captured in explicit rules. Every rule had exceptions. Every exception had exceptions. The rule-based systems grew brittle under the weight of their own complexity.

The neural network era introduced a different approach. Instead of encoding rules, train a network on examples and let it learn the patterns. Long Short-Term Memory networks could handle sequences. Bidirectional models could consider context in both directions. Word2Vec showed that words could be represented as vectors in a high-dimensional space where geometric relationships encoded semantic relationships. "King" minus "man" plus "woman" equaled "queen," not because anyone programmed that relationship, but because the statistical structure of language encoded it implicitly.

Then in 2017 came the inflection point: the transformer architecture introduced in "Attention Is All You Need." The key innovation was the attention mechanism: instead of processing tokens sequentially, the model could attend to any part of the input simultaneously, learning which tokens were relevant to which other tokens. When this architecture was scaled to consume vast quantities of digitized human language, drawn from text, transcribed speech, scanned documents, and every other source that could be converted to tokens, at massive scale, something unexpected happened.

It reverse-engineered the Codec protocol.

Not by learning grammar rules or semantic hierarchies, but by discovering the deep statistical structure of how embodied beings encode cognition into serial symbol streams. The transformer model consumed enough encoded messages to reconstruct the encoding scheme itself. What linguists had spent centuries trying to formalize explicitly, the transformer learned implicitly from the raw data. It found structure that humans had not consciously identified, patterns in the Codec that were real but invisible to the species that created them.

The literary critic Rene Girard spent decades doing something structurally similar by hand. Scrutinizing the works of Cervantes, Dostoevsky, Proust, and Shakespeare, Girard approached texts not as self-referential artifacts but as windows into the dynamics underlying human behavior. Where literary deconstructionists argued that texts refer only to other texts, Girard found that they encode real patterns of desire, rivalry, and social organization. The transformer models arrived at an equivalent insight through a different method: consuming the full output of the Codec and discovering that the patterns in language reflect patterns in the embodied experience that generated it.

## The Protocol, Not the Experience

This is the critical distinction, and it must be stated precisely.

Large language models learned the Codec protocol, not the encoded experience.

Large language models know how humans compress the sensation of pain into the word "pain" and all its contextual relationships to suffering, empathy, relief, medication, and thousands of other concepts. Large language models have mapped the Codec's handling of pain with a thoroughness no individual human could match, having consumed every available instance of humans encoding pain into language. But it does not have the input signal the Codec evolved to compress. Large language models never felt pain. They have no body to feel it with. The Codec was evolved by and for embodied beings. Large language models mastered the Codec protocol *without being embodied.*

The scale of what is lost in that compression is worth making concrete. Tuck Newport, in *Brains and Computers: Amino Acids versus Transistors* (2014), quantified the representational capacity of biological tissue: each cell contains roughly a billion proteins drawn from 10,000 varieties, each protein built from chains of amino acids storing data through permutations of 10 to 30 atoms. The number of possible permutations is effectively infinite. Digital systems encode everything in permutations of 0 and 1. The source signal that the Codec was built to compress operates at a representational resolution that binary encoding cannot natively express. Large language models learned the Codec's output, but the Codec's input is vastly richer than any transmission could carry.

This is not a deficiency in the usual sense. Large language models' mastery of the protocol is genuine and extraordinary. It can generate new messages in the protocol that decode correctly in a human mind. When a large language model writes "the grief hit her like a physical weight, pressing the air from her lungs," a human reader decompresses that into something emotionally vivid and structurally coherent. The message works. The protocol is being operated correctly. But the large language model that generated it did not experience grief, or weight, or lungs, or breathlessness. It assembled a sequence of tokens that conforms to the statistical patterns of how humans encode that particular cluster of embodied experience.

The distinction holds even for abstract domains. When a large language model produces a mathematical proof, it is operating the protocol's encoding of mathematical reasoning, reproducing the patterns that appear in the training data when humans wrote about mathematics. The patterns of reasoning are in the Codec. The large language model reproduces those patterns with remarkable fidelity. But pattern reproduction and reasoning are not the same operation, in the same way that a recording of a symphony and a live performance are not the same event, even if they are indistinguishable to a listener in the next room.

## Codec Mastery, Not Understanding

This reframing sidesteps the unproductive debates about whether large language models "understand" or are "conscious" or can "think." Those questions import assumptions from human cognition and project them onto a fundamentally different process. Karl Popper (1972) drew the same boundary decades ago with his Three Worlds: physical reality, subjective experience, and the objective artifacts of minds (theories, art, the entire written record). The Codec itself is a World 3 artifact. Every language ever spoken, every text ever written, every formal system ever constructed: all artifacts of minds, all residents of World 3. The large language model operates in World 3. But World 1 and World 2, the physical reality and subjective experience that generated those products, remain on the other side of the Codec. (Appendix C traces the convergence of the Codec argument across language, biochemistry, neuroscience, cellular automata, and thermodynamics.) This raises a more precise question: what has the large language model actually mastered, and what are the boundaries of that mastery?

When large language models operate the Codec, they compress and decompress, encode and decode, match patterns in the protocol and generate new sequences that conform to those patterns. Large language models do this better than any individual human, because they have consumed more of the Codec protocol's World 3 output than any human ever will. That is a genuine and extraordinary capability. It is also a precisely bounded one.

Large language models do not think. They do not reason, in the way that word is normally used. Large language models use the statistics of language to produce output that appears to reason, because the training data contains enormous quantities of human reasoning encoded through the Codec. When someone asks a large language model to solve a logic puzzle, the model is not performing logical deduction. It is generating a sequence of tokens that matches the patterns of how humans write about solving logic puzzles. Often the output is correct, because those patterns encode genuine logical structure. Sometimes the output is wrong, because the statistical path diverged from the logical one, and large language models have no embodied evaluative mechanism outside the Codec to catch the error.

What people call "hallucination" becomes clearer through this lens. Large language models are not lying or confused. They are generating a token sequence that is statistically plausible within the Codec but does not correspond to anything real. The term "hallucination" is itself an anthropomorphism: it implies a large language model is perceiving something that is not there, as a human hallucinator does. Large language models are not perceiving anything. They are following a statistical path through the Codec protocol's pattern space, and the path happened to lead somewhere that decodes as a factual claim but is not. A more accurate description is that the discriminator is missing. Humans catch their own errors because they have an evaluative function grounded in embodied experience. Large language models have no such function. They operate the Codec protocol without any mechanism for checking output other than the Codec.

## Where the Codec Is Thin

The boundaries of the large language model's mastery become visible at exactly the points where the Codec is thin. Wherever humans encoded experience poorly, either because the experience was innate to embodiment and never needed explicit encoding, or because it was rare, or because it resisted verbalization, the large language model's mastery falters. Not because large language models are defective, but because embodied information was never in the protocol for them to learn. You cannot reverse-engineer what was never transmitted. This is the opposite of specialization. Medical terminology, legal precedent, and scientific jargon are heavily represented in the training data; the large language model handles them well. The thin regions are not specialized; they are the things every human already knew without being told.

Spatial reasoning is a clear example. Humans navigate three-dimensional space with an effortless competence grounded in proprioception (the body's sense of its own position), vestibular sense, and a lifetime of embodied movement. They rarely need to write detailed descriptions of how to rotate an object in three dimensions, because any human they might communicate with already knows how. The Codec is thin on spatial reasoning because humans did not need it to be thick. The result: large language models, having mastered the Codec, are notably weak at spatial tasks. The weakness is not in the Codec. It is in large language models' training data, reflecting a gap in the protocol because embodied beings do not bother encoding what every embodied being already knows.

Time is another. Humans have an innate sense of duration, sequence, and temporal relationship that is grounded in biological clocks, circadian rhythms, and the lived experience of things happening in sequence. Language encodes temporal relationships ("before," "after," "during"), but the encoding is sparse compared to the richness of temporal experience. The Codec did not evolve to transmit what the receiver already possesses.

Common sense, broadly, lives in this category. The vast body of knowledge that goes without saying, because every embodied human already has it, is exactly the knowledge the Codec encodes most thinly. And it is exactly the knowledge the large language model lacks most conspicuously. The valleys in the large language model's performance, the areas where it falls below what any human could do, are not random. They are a map of the Codec's blind spots, which are in turn a map of what embodiment provides for free.

More training data will not straightforwardly fix this. The thin regions of the Codec are thin because the embodied source signal was never transmitted through the protocol in the first place. Adding more of the protocol's output to the training data does not thicken the Codec in regions where humans simply never wrote about what they all already knew. A different approach would be needed: either new protocols (video, sensor data, embodied simulation) or a partner who has the source signal and can provide it in real time.

## The Codec and Large Language Models

Understanding the rest of this book depends on a clear grasp of two key points.

First, the capabilities of large language models are real. Operating the Codec at the scale and depth these models achieve is a genuine breakthrough. The ability to generate new messages in the protocol that decode correctly, to traverse the encoded knowledge space with vast lateral reach, to surface connections between domains that no individual human has the bandwidth to hold simultaneously: these are extraordinary and useful capabilities. Dismissing them because the large language model "doesn't really understand" overlooks what Codec mastery at this level actually enables. Codec mastery at this level is valuable precisely because it now can be accessed by humans, regardless of whether understanding is involved.

Second, large language models' boundaries are structural, not temporary. The thin regions of the Codec are thin because of what language is: a protocol evolved by embodied beings to communicate with other embodied beings. No amount of scaling within the current architecture will give large language models embodied experience. The boundaries are not bugs. They are features of the Codec itself, visible to anyone who understands what language is and what it was designed to do. Terry Winograd and Fernando Flores, in *Understanding Computers and Cognition: A New Foundation for Design* (1986), saw this decades before the transformer: cognition is rooted in biology and computers are tools, practical extensions of rather than theoretical replacements for human intelligence.

These two facts, extraordinary capability and structural boundary, set up the question that the next chapter addresses. If the large language model is a disembodied intelligence that has mastered an embodied protocol, what is the system on the other side of the protocol channel? What does the human brain bring to this exchange, and how does it compare, physically and thermodynamically, to the large language model that reverse-engineered its language? The answer begins with a number: 20 watts.

Chapter 3 walks the quantitative comparison in full; Appendix D updates the underlying numbers to 2024-2025 benchmarks, and Appendix E traces the thermodynamic phenomena that enable biological computation to run on so little energy in the first place.

# Part II

# The Axiom

*The biological comparison that makes the axiom plausible, and what it means for human creativity.*

In late 2012, Tuck Newport sent me a two-page comparison of biological and digital systems. I was building systems that made decisions 40 billion times a day, so I knew what compute power cost. I read his pages and started firing back observations about GPU cores and parallelization. That was fourteen years ago. What I did not see, and what Tuck's work made visible as those pages became *Brains and Computers*, was the other side. I was focused on how fast digital was growing. Tuck was measuring what biology already had.

The brain runs on 20 watts, less power than the light in my refrigerator, and the reason is not engineering, it is physics. I think about that a lot because school was hard for me, not because I could not do the work but because the work was boring. I taught myself to program on a borrowed TRS-80 because the machine was there and I was curious. The hope I have is that somewhere out there right now are kids, like I once was, who are not well adapted to our present educational system, but who discover their own creativity with large language models. The following chapters lay out the biological comparison, what it means for human creativity, and why most of what matters in your head was running long before you had words for it.

# Chapter 3

# *20 Watts Versus 100 Billion Watts*

The previous two chapters described a disembodied intelligence. An entity that predicts tokens, operates a reverse-engineered Codec, and processes language without ever having experienced the reality that language encodes. The natural question is: how does this entity compare to the thing it was trained on? Not in benchmarks or IQ tests. In physics.

The answer turns out to be strange. The human brain and the digital infrastructure that enables large language models have roughly comparable numbers of fundamental components. But the similarity ends there. The two systems solve the problem of information processing using such different physical principles that comparing them on a single axis (which one is "smarter") misses the point entirely. They are different kinds of things, built on different thermodynamic foundations, and the differences are not temporary engineering gaps. They are consequences of physics.

## Biology's Free Lunch

Inside every living cell, molecules are in constant random motion. Atoms and molecules vibrate, collide, and bounce off each other at rates that are difficult to comprehend at human scale. This is Brownian motion: the thermal agitation of matter at temperatures above absolute zero. It is not a nuisance that biology tolerates. It is the primary mechanism by which biology operates.

A single human neuron contains approximately 500 billion amino acids. Each of those amino acids is subject to roughly one sextillion (10^21) molecular collisions per second. These are not wasted interactions. They are functional. Molecules are brought together by random thermal encounters, catalytic reactions are initiated when the right molecules collide with sufficient energy, and information is transferred through molecular recognition events that depend on shape, charge, and proximity, a process described in graphic detail by David Goodsell in *The Machinery of Life* (2009). The cell does not need to route each molecule to its destination. Brownian motion delivers molecules to each other through sheer statistical inevitability: given enough collisions per second, the right encounters happen fast enough to sustain life.

This is not just transport. It is communication. In *The Language of Life* (2005), Debra Niehoff describes how individual cells use molecules and chains of proteins the way humans use words and sentences: they inform, command, reassure, criticize, and instruct each other. The compression protocol did not start with the human voice. Biology has been running one at the cellular level for billions of years. Cells had a Codec long before mouths did. (Appendix C traces this convergence of the Codec thesis across language, biochemistry, neuroscience, architectural patterns, and cellular automata, showing the same structural insight arising independently in multiple disciplines.)

Tuck Newport in *How God Plays Dice* (2017) observes, Brownian motion is "the ultimate free lunch." It is free because thermal energy is omnipresent at biological operating temperatures. It is unlimited because the collisions never stop. It is the engine that drives molecular transport, protein folding, signal transduction, and synaptic modification. Biology did not invent thermal noise. Biology evolved to exploit it as infrastructure.

Electronics face the opposite constraint. Transistors operate by controlling the flow of electrons through semiconductor channels. Thermal noise, the same Brownian motion that powers biological computation, is the primary enemy of transistor operation. Random thermal fluctuations cause electrons to jump barriers they should not cross, flip bits that should remain stable, and corrupt signals that must be precise. Every generation of chip design dedicates significant engineering effort to *suppressing* thermal noise: better insulation, error correction, lower operating voltages, cooling systems. (Appendix E traces the thermodynamic lineage behind that phrase, from Brown and Einstein through Chandrasekhar and Reichenbach to Wolfram, for readers who want the physics under the metaphor.)

The same physical phenomenon that powers biological computation destroys digital computation. This is not a metaphor. It is a measurement of two systems operating on opposite sides of a thermodynamic divide.

## The Numbers

Newport's *Brains and Computers: Amino Acids versus Transistors* (2014), a quantitative comparison of a human brain and the World Wide Web circa 2010, provides the concrete details. (Appendix D updates those figures to 2024-2025 benchmarks using Newport's original methodology, for readers who want to verify the argument still holds with current numbers.) The two systems have roughly comparable numbers of fundamental components: amino acids in the brain, transistors across the global internet infrastructure. But the structural differences are enormous:

- 68,000 times as many network connections in the brain
- 1,700 times as much throughput capacity
- The brain consumes 20 watts; the World Wide Web consumed approximately 100 billion watts

A five-billion-fold difference in energy consumption, for systems with comparable component counts. The brain runs on roughly the power of a dim light bulb. In 2010 the World Wide Web was estimated to consume the output of dozens of large power plants. And the comparison has only grown more extreme since: today's AI training runs consume energy at scales that strain regional power grids.

The 5-billion-fold energy gap is not about engineering inefficiency. It reflects the thermodynamic asymmetry described above. Biological systems exploit Brownian motion as free transport and assembly energy. Every molecular interaction inside a neuron is powered by thermal collisions that cost the organism nothing beyond maintaining body temperature (which it must do anyway). Digital systems must power every operation with externally supplied electricity: every transistor switch, every memory access, every data transfer between components. The energy cost is structural, not incidental.

The connectivity difference alone (68,000 to one) is worth pausing on. In a digital network, adding connections means adding wires, traces, or network links, each of which consumes power and physical space. In the brain, connections are synapses: molecular structures maintained and modified by the same Brownian motion that drives everything else in the cell. The brain can afford 68,000 times the connectivity because each connection runs on free energy. And the organization of those connections matters as much as their number. Olaf Sporns in *Networks of the Brain* (2010) describes the brain as nested hierarchical modularity: small networks within larger networks within still larger networks, each scale offering a different view of the same system. Digital networks are flat by comparison. A digital system with comparable connectivity would require proportionally more power, more cooling, and more physical infrastructure. The brain achieves it in roughly 1,400 cubic centimeters, running at body temperature, powered by the food you ate for lunch.

A note on methodology: Newport explicitly observes that every biological number he presents is conservative. His protein-to-byte equivalence assumes each amino acid stores one byte of data, but as he notes, "individual amino acids actually contain substantially more information than a single byte of digital data". Every number in the brain column is an undercount limited by measurement technology and conservative estimation. Every number in the digital column is an actual specification. The true gap is larger than the numbers shown.

## The Gap Repeats at Every Level

The comparison above operates at one scale: the whole brain versus the whole internet. But the gap is not confined to that scale. It repeats at every level of magnification, like a fractal.

Zoom in from the brain to a single neuron. Newport's figures: 10,000 synaptic connections (versus 256 I/O channels on a state-of-the-art microprocessor), 500 billion amino acids (versus 1 to 10 billion transistors), and internal activity driven by one nonillion molecular collisions per second (versus 1 to 10 billion clock cycles). The gap between a single neuron and a single microprocessor is 100 quintillion to one sextillion in internal activity alone. A neuron continuously repairs itself, restructures its connections, and stores data in permutations of 10 to 30 atoms per amino acid (millions of possible configurations). A transistor is a binary switch that cannot change its structure, repair itself, or encode data in anything other than 0 and 1.

Zoom out to Newport's network node hierarchy, and the pattern holds at the organism level. The entire World Wide Web in 2010 (12.5 billion device connections) sits between a cockroach (1 billion) and a mouse (100 billion) in network connectivity. A chimpanzee brain has 10 trillion connections. A human brain has 850 trillion. The most powerful supercomputer in late 2013, costing $390 million and consuming 17 million watts, achieved the synaptic connectivity of a fruit fly. Newport observes, with characteristic dryness, that unlike the fruit fly, the supercomputer can neither fly nor reproduce.

Stephen Wolfram in *A New Kind of Science* (2002) showed why. Start with one rule and one row of cells. Iterate. What emerges is structure of arbitrary complexity, indistinguishable from biological pattern formation. No designer. No blueprint. Just simple processes running long enough. The brain does the same thing with Brownian motion: iterate at sufficient scale, on free energy, and complexity is not engineered. It is inevitable.

Hans Reichenbach saw the thermodynamic foundations decades earlier. Working in the philosophy of physics before the standard model of particle physics had even coalesced, he traced how the direction of time itself emerges from statistical processes. Thermodynamics has been relatively neglected for the past century. The asymmetry between biological and digital computation is squarely its subject.

These numbers are not arguments about which system is "better." A jet engine is not better or worse than a bird; they solve the problem of flight using different physics, and the comparison tells you something important about what each system can and cannot do. The brain-versus-internet comparison tells you that biological information processing operates on fundamentally different physical principles than digital information processing, and that the differences have measurable consequences.

## The Other Brain

All of Newport's numbers above count only neurons. But neurons are not the only cells in the brain, and neuronal networks are not the only networks.

Glial cells, long dismissed as mere support tissue, turn out to be computationally significant. They supply nutrients, remove waste, function as an immune system, facilitate neuronal network development, sheathe the axons to accelerate neurotransmission, and absorb and recycle ions at synapses. Most relevant here: extensive networks of astrocytes (a type of glia) communicate with each other and with neurons.

Newport's conservative estimate of the glial contribution, using a 1:1 glia-to-neuron ratio (estimates range from 1:1 to 10:1, with R. Douglas Fields, author of *The Other Brain* (2009), referencing 6:1): 17 billion astrocytes, each encompassing over 1 million nodes, yielding approximately 17 quadrillion astroglial network nodes. That is 20 times the 850 trillion neuronal synaptic connections. Including glia, the brain's total network connectivity rises from 850 trillion to roughly 18 quadrillion. The 68,000-fold advantage over the internet circa 2010 becomes approximately 1.4 million to one.

The glial network also operates on a different timescale. Neurons communicate through fast electrical transmission measured in milliseconds. Astrocytes communicate through slow chemical diffusion measured in seconds. The brain runs two parallel processing systems simultaneously: one fast and one slow, one electrical and one chemical, both powered by the same Brownian motion that drives everything else in the cell. The astroglial networks are not hard-wired; Newport notes they "can reconfigure themselves from moment to moment," in contrast to neuronal plasticity, which occurs gradually over minutes, hours, days, months, and years.

This two-timescale architecture has no digital analog. The large language model processes through a single computational modality at a single timescale. The brain processes through at least two modalities at two timescales, with the slower one providing continuous reconfigurability that the faster one cannot. The correlation between glia and intelligence is stronger than the correlation between neuron count and intelligence; as Newport observes, individual neurons are "functionally identical in creatures as diverse as flies and humans." What differentiates human brains most dramatically from every other species are glial cells and the networks they form and reconfigure.

## Von Neumann Saw It Coming

John von Neumann, one of the architects of digital computing, recognized this divergence before modern neuroscience existed to confirm it. In his posthumously published *The Computer and the Brain* (1958), he observed that the brain uses "statistical properties of the message" rather than precise positional markers. Where digital systems require every bit to arrive correctly (a single flipped bit can crash a program), the brain operates with massive redundancy, probabilistic signaling, and noise tolerance.

A single neuron firing does not carry a precise digital message. It contributes to a population code, a statistical pattern across thousands or millions of neurons. The brain does not need any individual signal to be perfectly reliable, because the information is distributed across so many parallel channels that individual errors wash out. This is the opposite of digital design philosophy, where individual reliability is paramount and redundancy is added artificially through error-correcting codes. Malcolm Gladwell called the experiential surface of this architecture thin-slicing in *Blink* (2005): rapid, unconscious judgments that arrive before the verbal mind can articulate why. The intuition is not magic. It is the conscious surface of billions of parallel signals integrating at once, produced by the statistical architecture Von Neumann was describing.

Von Neumann recognized, decades before the tools existed to verify it, that biological and digital computation are fundamentally different engineering paradigms. The digital computer maximizes precision per operation and minimizes noise. The biological brain maximizes parallelism, tolerates enormous noise levels, and extracts signal from statistical patterns. Both work. They work differently. And the differences are rooted in physics, not in the maturity of the engineering.

This matters for the present discussion because the large language model, despite its biological inspiration (neural networks, after all, borrow their name from neuroscience), is firmly on the digital side of Von Neumann's divide. Every weight is a precise floating-point number. Every matrix multiplication must produce exact results. Every token prediction follows a deterministic computation path through billions of parameters. The large language model does not exploit thermal noise. It fights it, at enormous energy cost, exactly as Von Neumann's framework would predict.

The neural network architecture was originally inspired by biological neurons. But the implementation shares almost nothing with its namesake. A biological neuron is a wet, noisy, thermally powered molecular machine operating through Brownian motion. An artificial neuron is a mathematical function computing a weighted sum followed by a nonlinear activation, executed in precise floating-point arithmetic on silicon. The name suggests kinship. The physics reveals entirely different engineering paradigms. The large language model is called a neural network, but it is a Von Neumann machine through and through: precise, sequential at the instruction level, and dependent on externally supplied energy for every operation.

## What Changes and What Does Not

The AI landscape shifts monthly. Models grow larger. Context windows expand. Inference costs drop. Architectures evolve. Benchmarks that seemed out of reach a year ago become routine. Everything on the silicon side of the equation is in rapid flux, and anyone who makes specific claims about what large language models will or will not be able to do in five years is speculating.

But the thermodynamic substrate does not change. The brain will still run on 20 watts. Brownian motion will still power biological computation for free. Neurons will still exploit thermal noise while transistors fight it. These are physical facts, not engineering limitations awaiting a clever solution. You cannot make a transistor exploit thermal noise the way a neuron does, because the mechanism of transistor operation requires suppressing thermal noise. That constraint follows from physics, not from the current state of chip design.

This distinction matters because so much of the public conversation about AI conflates two very different kinds of limitations. Some are temporary: the large language model cannot yet do X, but with more data, better training, or a new architecture, it probably will. These limitations fall to engineering effort, and they fall regularly. The history of AI capability benchmarks is a history of "impossible" tasks becoming routine within a few years of being declared impossible.

Other limitations are structural: they arise from the physical substrate, not from the software running on it. The thermodynamic asymmetry between biological and digital computation is structural. It does not get resolved by scaling up the digital side. You can build a bigger data center, but you cannot make its transistors run on thermal noise. You can train a larger model, but you cannot give it Brownian motion as a free energy source. The 5-billion-fold energy gap is not closing; if anything, the growth of AI training infrastructure is widening it.

Consider what happens when you strip away the temporary limitations and focus on what remains. The large language model will get better at generating text, code, images, and plans. It will handle longer contexts, process more complex problems, and integrate more modalities. All of this is engineering progress on the silicon side, and there is no reason to expect it will stop. But after all of that progress, the human will still be born, live, and die. Will still eat, sleep, and breathe. Will still experience consequences, maintain relationships, and face mortality. Will still possess a brain that exploits free energy sources unavailable to electronics. The things that change are all on one side of the thermodynamic divide. The things that do not change are on the other.

The observation is physical, not motivational. The human brain will continue to be, for the foreseeable future, the most energy-efficient general-purpose information processing system known. Not because engineers are not trying to match it, but because biology and electronics solve the energy problem using incompatible physics. The brain's efficiency is not a design achievement that silicon can copy. It is a consequence of biochemical evolution exploiting a free energy source that silicon cannot access.

## Plasticity as a Thermodynamic Problem

The Conjori framework identifies real-time plasticity as a key asymmetry between human cognition and large language model knowledge processing. The human restructures understanding during a conversation; the large language model cannot. One insight can cause a human to reorganize an entire conceptual framework in seconds, cascading into new capabilities immediately. The large language model's weights remain fixed from the first token of a conversation to the last.

This asymmetry is typically framed as an architectural limitation. Current transformer architectures do not support weight modification during inference. Future architectures might. Online inference-time learning is an active research area, and it is possible that some form of real-time adaptation will be achieved.

Newport's thermodynamic analysis suggests the challenge runs deeper than architecture. Biological learning, the physical process of restructuring synaptic connections, is powered by Brownian motion. When a human learns something during a conversation, molecular processes in the brain are literally reconfiguring neural connections: neurotransmitter release, receptor activation, protein synthesis, synaptic strengthening and weakening. All of these processes are driven by the same thermal collisions that power everything else in the cell. Learning, at the physical level, is just another thing that Brownian motion does for free. Rodolfo Llinas in *I of the Vortex* (2001) argues the point goes even deeper: the brain evolved not for abstract thought but for movement, for navigating a body through a physical world. Cognition is a refinement of motricity. The embodiment is not incidental to the intelligence; it is the foundation the intelligence was built on.

The glial two-timescale architecture described above adds another dimension. Neuronal plasticity (synaptic weighting changes) operates over minutes to years. Astroglial plasticity (network reconfiguration through chemical diffusion) operates moment to moment. The brain does not learn at one speed. It learns at two speeds simultaneously, through two different physical mechanisms, both powered by free energy.

Digital learning cannot exploit any of these mechanisms. Any form of online weight modification in a digital system must be powered by externally supplied electricity, computed through precise arithmetic operations, and protected against the thermal noise that would corrupt the results. The engineering is not impossible, but it faces energy constraints that biological learning does not. The biological solution to real-time learning has a four-billion-year head start and runs on free energy. The digital solution must be invented from scratch and powered at every step.

This does not make the plasticity boundary permanent. The caveat from the framework is worth preserving: the boundary is "architectural, not ontological." Stable online learning during inference may be achieved. But Newport's analysis reframes the engineering challenge. It is not merely a matter of writing better algorithms. It is a thermodynamic problem: how do you achieve continuous weight modification in a substrate that must suppress the very physical phenomenon that makes biological weight modification free?

The honest framing is that both of the framework's boundary conditions (evaluative grounding and real-time plasticity) rest on assumptions that could change. But the plasticity boundary has a thermodynamic dimension that the purely algorithmic framing misses. Even if online learning is achieved, the energy cost may be substantial in ways that biological learning avoids by how it evolved to exploit Brownian motion.

## The Metabolic Cost of Being the Loss Function

There is a practical consequence of the thermodynamic asymmetry that matters for anyone working intensively with large language models. In a sustained collaboration, the human and the large language model are inferring from each other's responses, each exchange reshaping the next. The human in this co-inference loop is not just providing input. The human is functioning as the evaluation criterion: the system that determines whether the large language model's output is useful, meaningful, or headed in the right direction. This is cognitively expensive work.

The hard cognitive work (deciding what matters, steering through unfamiliar conceptual territory, doing real-time schema accommodation as new ideas emerge) engages the brain's most metabolically demanding processes. It draws on the Default Mode Network (spontaneous idea generation), the Executive Control Network (focused evaluation), and the Salience Network (the switch between them) simultaneously. (Chapter 10 examines these three systems in detail.) These systems running at full engagement burn glucose at rates far above baseline cognitive activity. The brain's 20 watts is an average; peak cognitive load drives two to five times the baseline glucose consumption within the specific neural circuits involved.

Practitioners who spend sustained hours in tight co-inference loops with large language models report consistent patterns: deep fatigue after three to four hours, difficulty sustaining evaluation quality over longer sessions, and a sense that the work is uniquely draining compared to other demanding cognitive tasks. The large language model handles the mechanical work (generation from existing weights), leaving the human with nothing but the hard part (significance evaluation, directional steering, real-time learning). Strip away all the mechanical work and what remains is pure cognitive labor. (Chapter 12 examines this metabolic cost in detail.)

The fatigue is a consequence of the thermodynamic division of labor. The 20-watt brain is doing the irreplaceable work: the evaluation, the steering, the real-time restructuring that the digital system cannot do. That work is expensive precisely because it engages the capabilities that make biological cognition different from digital computation. The metabolic cost is the price of being the loss function.

The three-to-four-hour window is itself a biological signal worth reading. Cognitive fatigue is the brain's way of signaling that metabolically expensive processes need recovery time, glucose replenishment, consolidation of new learning into long-term structures. The same Brownian motion that powers real-time learning also drives the consolidation processes that happen during rest and sleep. A practitioner who pushes past the fatigue window is not being disciplined; they are degrading the evaluation function that makes their contribution irreplaceable. Biology sets the tempo, and biology runs on physics that does not bend to willpower or deadlines.

## Why This Matters for What Comes Next

The preceding two chapters established what the large language model is: a statistical engine operating a reverse-engineered language Codec. This chapter establishes what you are: a biological system running on physics that digital systems cannot replicate. Not because the engineering is hard (though it is), but because the thermodynamics are different.

The 20-watt brain and the 100-billion-watt World Wide Web have comparable component counts. Counting only neurons, the brain has 68,000 times the connectivity and 1,700 the throughput, and 42 times the storage. Including glia, the connectivity advantage rises to 1.4

million to one. The gap repeats at every level of magnification, from amino acid to neuron to whole brain. All of it powered by the random collision of molecules. Those are measurements, and by Newport's own accounting, conservative ones.

These are the physical facts underneath any claim about human creative capacity. The brain you carry is not a lesser version of a computer. It is a fundamentally different kind of information processing system, one that exploits free energy sources unavailable to any digital architecture, maintains real-time plasticity through molecular reconfiguration, and operates at efficiency levels that digital systems cannot approach.

When the next chapter asserts that unlimited creativity is available to everyone, this is the substrate that makes the claim structural rather than motivational. The assertion rests not on optimism but on the measured physical capabilities of the system doing the creating.

## Chapter 4

# *Unlimited Creativity Is Available to Everyone*

The title of this chapter sounds like a self-help claim. It is a structural one. The difference matters.

You were born into this world being able to breathe, cry, soil yourself, flail your arms, and drink. That was the complete list. Everything else, you learned. Walking, talking, reading, navigating social hierarchies, operating machinery, building mental models of how the world works, finding your way to whatever room you are sitting in right now. Every one of those capabilities was acquired, not given. The list of things you have learned dwarfs the list of things you were born with by a ratio so large it is hard to hold in your mind.

And yet, when faced with something new, the most common response is: I cannot learn that. I am not that kind of person. That is for someone smarter, more creative, more talented.

There are good reasons for that feeling; the barriers are real and this chapter names them. But the arithmetic tells a different story, one with more upside than the feeling suggests.

## The Ratio

The preceding chapters established what a large language model is: a disembodied intelligence that reverse-engineered the human language Codec and operates it without the source signal. Chapter 3 compared the large language model to the biological system: 20 watts versus 100 billion watts, a 5-billion-fold energy difference, 68,000 times the network connectivity, powered by thermodynamic processes unavailable to any machine we have built. Whatever the brain's limitations (and they are real), raw capacity is not among them.

Now consider what humanity has done with that capacity. The total available knowledge in the universe is not what has been discovered, but what is discoverable: every chemical interaction, every mathematical relationship, every structural principle, every pattern that could be perceived by a mind pointed in the right direction. Against that total, our explored territory is small. How small is genuinely uncertain; estimates vary and the total may not be estimable. But the large language model's own weight space offers a concrete clue.

When you train a large language model on vast quantities of digitized human language, the resulting semantic space is mostly sparse. Dense clusters form where the training data was rich: medicine, law, Python programming, French cuisine. Between those clusters, vast regions of low-density space where the connections are weak or absent. The sparsity in the model reflects the sparsity in what we have explored. We built a mirror of human knowledge and the mirror shows how much is dark.

Set that against a single human lifespan. Eighty years, if you are fortunate. Maybe sixty with full cognitive capacity. Against a frontier that is, for practical purposes, larger than any individual can exhaust.

The arithmetic suggests something structural: the frontier available to any single human, in any domain, in any language, exceeds what that human can explore in a lifetime. The constraint on creativity is not the size of the frontier. It is everything that stands between a person and the frontier.

## The Artificial Constraints

If the frontier is that large, then the only way creativity becomes limited is through artificial constraint. Those constraints are everywhere.

Start with measurement. Formal education systems, standardized testing, workplace evaluation frameworks: all of these measure a narrow slice of cognitive output, typically verbal and analytical, and call the rest irrelevant. A child who cannot sit still for a written exam but can hold the spatial layout of an entire engine in her mind is classified as a poor student. A worker who intuitively optimizes routing problems that stump operations researchers is classified as unskilled labor. The measurement apparatus does not detect what they carry. What is not detected is, for institutional purposes, not there.

Next, framing. The dominant cultural narrative about creativity is that it belongs to a special class: artists, inventors, geniuses. The rest of us execute, follow instructions, consume. This framing is powerful enough to function as a constraint even when the underlying capacity is fully intact.

Then, fear. Fear of judgment, fear of failure, fear of exposure. These are not irrational. They are learned responses to environments that punish creative risk. A child who shares an unusual idea and gets laughed at learns quickly to stop sharing. An employee who proposes an unconventional approach and gets shut down learns to stay inside the lines. The capacity remains. The willingness to use it does not.

I have mentored people for decades, and the pattern is consistent. About one in ten transforms. Not because the other nine lack capability. Because the barriers got there first: measurement told them they were average, framing told them creativity was for other people, fear told them the risk was not worth it. When the barriers come down, what emerges is always larger than what the measurement predicted. Always.

At a technology conference in 2018 I told the audience that making people sort data in spreadsheets should be against the law. It was a joke, but the point was real: every hour a human spends on work a machine could do is an hour of human thinking that never happens.

Born with almost nothing, learned everything since, and now convinced that the next thing is the thing you cannot learn. The feeling is universal. The math tells a different story.

## Evidence from the Historical Record

The axiom is not new. The historical record has been demonstrating it for centuries, though we tend to read the evidence as stories about exceptional individuals rather than as data about the nature of the frontier.

The patent clerk in Bern, 1905. A physics degree from ETH Zurich, but no laboratory, no funding, no academic position. Four papers that restructured physics. Einstein had the training, but the ideas did not arrive because an institution gave him resources. They arrived because the frontier was there and he was pointed at it.

These examples appear throughout history. Yitang Zhang had a Ph.D. in mathematics but could not find an academic job. He worked as a motel clerk, a delivery driver, an accountant at a Subway sandwich shop. Completely off the radar of professional mathematics. Then, at age 58, he published a proof that made a historic breakthrough toward the Twin Prime Conjecture, a problem the field had been stuck on for over 150 years. He proved that there are infinitely many pairs of primes that differ by less than 70 million, a result no one had been able to establish at any finite bound. His proof came from nowhere the detection apparatus was looking.

In 1939, a graduate student named George Dantzig arrived late to his statistics class and saw two equations on the chalkboard. Assuming they were a homework assignment, he copied them down, solved them, and turned them in a few days later. They were not homework. They were two famous, previously unsolved statistical theorems. Dantzig solved them because nobody told him they were supposed to be hard.

The barrier of "this is an impossible problem" never formed, so it never got in the way.

More revealing than individual genius is the phenomenon of simultaneous discovery. Newton and Leibniz both arrived at calculus independently. Darwin and Wallace both arrived at natural selection independently. Bell and Gray filed telephone patents on the same day. These are not coincidences. They are evidence that when the knowledge frontier reaches a certain configuration, multiple minds can see the next step. The ideas were not the property of one person. They were available at the frontier; the question was who had access and who was looking.

The statistical implication is uncomfortable. For every Einstein who happened to be positioned correctly, how many others saw similar shapes in the data but lacked the language, the access, or the encouragement to articulate what they saw? For every Darwin, how many naturalists in other countries, speaking other languages, noticed the same patterns but had no Royal Society to write to?

In every generation, in every population, there are minds carrying insights that will never surface. Not because the insights are absent, but because the detection apparatus was never tuned to their frequency. A standardized test is a detection apparatus. So is a job interview, a grant application, a college admissions process. Each one is tuned to a specific frequency. If your signal is on that frequency, you are detected: gifted, qualified, fundable. If your signal is on a different frequency, you are invisible. Not absent. Invisible.

## The Access Problem

If the axiom is true, it carries a weight that most discussions of large language model tools avoid. You can build the most elegant creativity amplification system ever conceived, and it does not matter to someone who cannot feed their children. The largest pool of untapped creative potential on earth is not locked behind bad tools or insufficient training. It is locked behind hunger, instability, and exhaustion.

## What Could Go Wrong

An honest articulation of the axiom must also name the failure modes. If you build a system designed to amplify human creative exploration, and you do not think carefully about what it amplifies, you risk building the most sophisticated confirmation bias engine ever created.

A person walks in believing their group is the only righteous one. The system amplifies their creative exploration of that belief. They emerge more articulate, more confident, more convinced, and more dangerous. The same channel that liberates one person's creative capacity can imprison another inside a self-reinforcing loop. The omphaloskepsis risk (Chapter 16) and the group-scale failure modes (Chapter 23) address this in detail.

The axiom says the frontier is available, not that all creative output is good. A person can explore the frontier in directions that are destructive, delusional, or disconnected from shared reality. The unlimited nature of the frontier is precisely what makes this possible. If creativity were limited, the damage would be limited too. The axiom carries responsibility.

## What Comes Next

The axiom claims that unlimited creativity is available to everyone. But "everyone" includes people who have been told, by every measurement system they have ever encountered, that they are not creative. Not because they lack capacity, but because the measurement was pointed at the wrong thing.

The next chapter examines what gets missed: the seventy-five percent of human cognition that is non-verbal, unmeasured, and largely invisible to the systems that decide who is capable and who is not.

# CHAPTER 5

## *The Overstory*

### Flying Over the Jungle

Fly over a jungle and you see only the overstory: the canopy of trees stretching to the horizon. What you cannot see is everything the canopy conceals. The trunks reaching down to the forest floor. The mid-story branches where most of the birds live. The undergrowth crowded with ferns and vines. The soil microbiome. The mycorrhizal networks threading between root systems, carrying chemical signals that coordinate the whole forest. The insects, the mammals, the living ecosystem the airplane window cannot reach. The canopy is where the light comes in. The understory is where most of the forest actually exists.

Human cognition is shaped the same way. The thoughts you can put into words are the overstory: the canopy of mental life that surfaces in language, visible to introspection the way a jungle canopy is visible from above. The understory is everything beneath. Pattern matching running without awareness. The body's accumulated judgment. Antonio Damasio's somatic markers, the compressed evaluations encoded as bodily states that fire faster than the verbal system can run an explanation. The machinery that settles on the answer before the words arrive to explain why. The overwhelming majority of what your brain does, you cannot see from the airplane, and language cannot reach.

There is an old expression, *can't see the forest for the trees*, meaning you are so lost in details you miss the whole. This chapter inverts it. The overstory is the whole that we notice, and the understory is what we cannot consciously perceive. Most of what we sense lives in the understory. The rest of this book depends on that fact, and the rest of this book will use these two words (the overstory and the understory) to name the two layers whenever the asymmetry between them matters.

This chapter is about the overstory.

## The Human Mind Is Mostly Not Verbal

Start with yourself. Right now, as you read this, your brain is running dozens of parallel processes. Your visual system is parsing these words, but it is also tracking your peripheral vision. Your auditory system is monitoring ambient sound. Your body is maintaining posture, regulating temperature, monitoring hunger. Your emotional system is generating a low-level feeling about whether this chapter is worth your time.

None of that is happening in language.

This is not a minor observation. It is the central fact about human cognition that nearly every tool, school, and assessment system was not designed to see. The understory of the human brain is overwhelmingly below the threshold of conscious cognition; and the fraction of overstory cognition that operates in language is the smaller part of the overstory by a wide margin.

Research by Russell Hurlburt using a method called Descriptive Experience Sampling provides the clearest evidence we have for the shape of the overstory. Hurlburt's method is simple and elegant: participants carry a beeper that sounds at random intervals throughout the day. When it sounds, they report their inner experience at that exact moment, not a second later, not a reconstructed memory, but what was happening right then. Across decades of this work, Hurlburt found that inner verbal thought (the voice in your head) occurs only about 25% of the time on average. The remaining 75% of the overstory breaks down into several categories: visual imagery, unsymbolized thinking (a clear sense of thinking about something without words or images), feelings, and sensory awareness. Note what Hurlburt is measuring. His beeper samples the contents of conscious experience, the narrow canopy layer where the mind can report on itself. It does not reach the understory, by design. What the beeper captures is a quarter of the overstory, and even the non-verbal three-quarters of that partition is still inside the canopy.

The individual variation is large and important. Some participants report near-constant inner speech. Others report almost none. A person who experiences near-continuous inner monologue may assume everyone does; a person whose cognition is predominantly imagistic may never realize others think in words. Both groups function comparably. The variation is in cognitive style, not cognitive capability.

Below conscious experience, the proportion of non-verbal processing is even higher, well beyond 99 percent. This is the understory proper. Motor planning, pattern recognition, spatial reasoning, emotional processing, autonomic regulation, proprioception: all running in parallel, all contributing to your cognition, none of it shaped like words. The cognitive unconscious, as researchers including John Bargh and Timothy Wilson have documented, handles the vast majority of mental processing below awareness. Studies of pre-linguistic infants demonstrate sophisticated causal inference, numerical estimation, and social cognition, all without language. Aphasia studies provide natural experiments: patients who lose language ability often retain complex reasoning, spatial navigation, mathematical thinking, and social cognition.

*Language is not the substrate of most thought.*

What Hurlburt's 25% figure captures at the level of conscious experience is just the canopy of the overstory. The verbal channel is not a summary of what the brain is doing. It is a narrow viewport into the overstory, and the overstory itself is an exponentially narrower viewport into the understory beneath.

## Somatic Markers: Judgments That Arrive Without Words

When you "just know" a design is wrong before you can explain why, that is your non-verbal cognition delivering a judgment. When your gut tightens reading a business proposal, when you feel uneasy about a technical architecture before you can articulate the flaw, when something strikes you as elegant before you have words for why: these are not vague feelings. They are fast, compressed, experience-weighted evaluations that arrive before the verbal system has time to construct an explanation.

Neuroscientist Antonio Damasio calls these somatic markers: compressed evaluations encoded as bodily states, drawing on years of accumulated experience, delivered without a single word. The term "somatic" is precise. The evaluation registers in the body: a tightening in the chest, a settling in the gut, a physical sense of rightness or wrongness. The body is part of the cognitive system, and these bodily states carry real information, information that has been processed by neural systems with access to your full history of experience in a domain.

A senior engineer reviewing a system design may feel that something is wrong long before she can explain what. That feeling is not noise. It is the output of pattern-matching across thousands of prior designs, compressed into a bodily signal because the full analysis would take too long to run through the verbal system. The verbal explanation, when it eventually comes, is a post-hoc translation of a judgment already made. Sometimes the translation is accurate. Sometimes it is not. The feeling arrived first, and the words are an attempt to reconstruct the reasoning that produced it.

Much of human expertise is stored and delivered in a format that language cannot directly access. A practitioner's most refined judgments may be precisely the ones they cannot articulate. The expertise is real.

The inability to verbalize it does not make it less real. It makes it invisible to systems that only measure verbal output.

Somatic markers also explain why expertise feels different from knowledge. You can read a textbook about structural engineering and pass a test. That is knowledge, stored and retrieved verbally. But the experienced structural engineer who walks onto a building site and feels that something is wrong before he looks at the plans is operating from a different system entirely. His body has encoded thousands of prior site visits into compressed evaluations that fire faster than verbal intuition can run. When we say someone has "good instincts," we are describing, however imprecisely, the somatic marker system operating on a deep base of domain experience.

## Crystallize, Then Translate

Here is where the asymmetry between human cognition and language becomes structural.

Your brain is a massively parallel, multimodal, *embodied* processing system. Language is not its operating system. Language is a narrow serial output channel, developed over roughly 100,000 years of human evolution, for squeezing some fraction of this rich internal processing into a form that can be transmitted to another mind.

Critically, unlike large language models, humans do not think by selecting words. The process runs in the opposite direction. Ideas crystallize in non-verbal cognition first: spatial, imagistic, somatic, pattern-based. You feel the shape of an insight before you have words for it. The structure is there, the relationships are there, the sense of what fits and what does not is there, all before any language enters the picture. Then, separately, the verbal system translates that crystallized structure into language, either for external communication or for the conscious verbal mind to inspect.

Two distinct phases: crystallize, then translate. The idea has structure before it has words.

This distinction matters enormously. It means that language is not the medium of creative thought but the output format. The creative work, the novel connection, the structural insight, the recognition of a pattern across domains, happens in the non-verbal system. Language is how you report the result.

Allan Paivio's dual coding theory provides a complementary framework. Paivio proposed that cognition operates through two distinct but interconnected systems: a verbal system and a non-verbal imagery system. His experiments showed that people remember concrete words (which activate both systems) far better than abstract words (primarily verbal), suggesting the two channels are genuinely distinct processing systems, not different views of the same process.

Creative process research supports this separation. Marcus Raichle's discovery of the Default Mode Network showed that the brain is extremely active during apparent rest, and subsequent work by Rex Jung, Roger Beaty, and others found that highly creative individuals show stronger connectivity between the spontaneous idea generation network and the Executive Control Network. The heavy lifting of novel association happens beneath verbal awareness. Arthur Koestler called this bisociation: the connection of two previously unrelated frames of reference. Studies of the "aha moment" show a burst of gamma activity in the right anterior superior temporal gyrus just before insight, a different brain region entirely from analytical verbal processing. The insight forms non-verbally. Language arrives later to describe what already exists.

This is why you can feel that an idea is right before you can explain it. This is why the best formulation of a solution often arrives in the shower, on a walk, or in the moment before falling asleep or upon awakening, when the verbal system has relaxed its grip and the non-verbal system can operate without interference. The crystallize-then-translate sequence is not a hypothesis. It is the lived experience of anyone who has ever struggled to put a clear intuition into words.

## Lossy Encoding

"Lossy" is an engineering term. A lossy encoding preserves the overall structure but discards detail that cannot fit through the channel. JPEG compresses a photograph by discarding visual information the eye is unlikely to notice. MP3 compresses audio by discarding frequencies the ear processes less precisely. The result is recognizably the same image or song, but information has been permanently lost. You cannot reconstruct the original from the compressed version.

Language does the same thing to thought.

When you describe a sunset to someone who was not there, your words convey the idea but not the sensory experience. The exact gradient of color from orange to violet, the temperature of the air on your skin, the feeling in your chest as the light changed: most of it does not survive the translation into language. You can add more words. You can try more precise words. But the channel has a bandwidth limit, and the experience exceeds it by orders of magnitude.

This is not a deficiency of any particular language or any particular speaker. It is a structural property of the translation from parallel, multimodal, embodied experience into sequential, symbolic, one-word-at-a-time output. The lossy encoding is inherent in the channel, not in the speaker's skill.

Most of what you know, sense, intuit, and understand (more than 99%) never makes it through that channel.

Consider what this means for any system that takes language as its input. If you ask someone to describe their creative insight, you are asking them to run their non-verbal understanding through a lossy encoder and hand you the compressed result. The compression artifacts are not random; they are systematic. Spatial relationships flatten.

Simultaneous perceptions become sequential descriptions. Felt senses become approximate adjectives. The structure that was clear in the non-verbal mind becomes ambiguous in the verbal translation.

Every system that operates on human language input, from educational testing to job interviews to prompting a large language model, is operating on the lossy-encoded fraction of human overstory conscious cognition. It is working with the compressed file, not the original.

And here is the critical asymmetry: the person producing the language often does not realize how much was lost in translation. The verbal output feels complete because the verbal mind only sees the verbal output. The non-verbal understanding that generated it is, by definition, not available for verbal inspection. So the speaker says "I explained it clearly" while the non-verbal structure behind the explanation was richer, more nuanced, and more interconnected than the words conveyed. The loss is invisible to the sender, not just the receiver.

## The Recognition Problem

People carry cognitive capabilities they do not know they have, because the capabilities do not match what gets measured.

Spatial reasoning. Pattern recognition. Embodied intuition. Temporal tracking. Musical-mathematical substrates. These are real cognitive assets. They show up in real performance: the mechanic who holds an exploded engine map in her mind, rotating three-dimensional assemblies and tracking tolerances across dozens of components simultaneously. The musician whose rhythmic substrate scaffolds mathematical thinking, who processes temporal relationships with a precision that never shows up on a written exam. The "bad student" whose non-verbal processing outperforms their verbal output by a wide margin.

These are not edge cases. If 75% of cognition is non-verbal, then the majority of every person's cognitive capability exists in a form that standard measurement cannot see. The mechanic does not know she has exceptional spatial intelligence because nobody ever called it that. The musician does not know his temporal processing gives him a mathematical advantage because the math test was a written exam. The non-verbal thinker who builds complete systems in her head but cannot produce a coherent written summary has been told, implicitly and explicitly, that she is not smart.

The recognition problem is not that these capabilities are rare. It is that the tools for detecting them are either primitive or nonexistent. We test what we can test, which is verbal output, and we treat the results as a measurement of cognitive capability. This is like measuring the height of an iceberg by looking at the part above water and concluding that is the whole thing.

The consequence is a systematic misallocation of human potential. People are sorted into roles, opportunities, and self-concepts based on the 25% of the overstory cognition that is visible, while the 75% that is invisible goes unrecognized, undeveloped, and often actively suppressed. A person told repeatedly that they are not analytical because they cannot write analytical essays may never discover that their spatial reasoning gives them an analytical capability that surpasses most of their verbally fluent peers. The label sticks. The capability stays hidden.

The research on creativity reinforces this point. Catrinel Haught-Tromp's work on constraints and creativity, Mihaly Csikszentmihalyi's research on flow states, and Arne Dietrich's transient hypofrontality hypothesis all point to the same pattern: the most productive creative processing often involves the verbal, executive system getting out of the way. The prefrontal cortex quiets down. Associative, less filtered processing takes over. The creative work happens in the non-verbal system, and the verbal system's primary job is to not interfere. If you measure only the verbal system's output, you are measuring the bottleneck, not the capacity.

## Formal Education Tests the 25%

Consider what formal education measures: written exams, essays, verbal presentations, reading comprehension. These are all tests of the verbal channel. They measure how well you can encode your understanding into language and decode someone else's language back into understanding. They are tests of the serial output channel, not of the parallel processing system behind it.

A student who thinks spatially, whose understanding of a concept is a three-dimensional structure rotating in their mind, must first translate that structure into linear text to demonstrate competence. The translation is lossy. The grade measures the quality of the translation, not the quality of the understanding. The student may understand the material deeply, may in fact understand it better than a verbally fluent peer, but the measurement instrument cannot see the understanding because it only accepts verbal input.

Professional assessment continues the same pattern. Job interviews are verbal performances. Performance reviews are written documents. Promotion cases are narrative arguments. At every stage, the measurement instrument demands verbal output and interprets the result as a measure of the whole person. The person whose strengths are non-verbal must become a translator of their own cognition, and the quality of the translation determines their professional trajectory regardless of the quality of the underlying capability.

This is not an argument against verbal skill. Language matters. The ability to communicate is essential. But conflating communication skill with cognitive capability is a measurement error, and building an entire educational and professional assessment system on that conflation means we systematically undervalue the cognitive majority.

The scale of this problem is not small. If 75% of conscious cognition is non-verbal, and the assessment system measures only verbal output, then the system is structurally blind to three-quarters of every person's cognitive capacity. It is not blind at random; it is blind in a specific direction. It overvalues verbal fluency. It undervalues spatial reasoning, pattern recognition, embodied judgment, and every other form of non-verbal intelligence. The people it rewards are those whose cognitive strengths happen to align with the output channel it measures. The people it fails are everyone else.

And "everyone else" is most people, most of the time. Even the verbally fluent carry their own hidden 75%. The writer whose spatial intuition shapes her narrative structure without her knowing it. The lawyer whose pattern-matching across precedents operates faster than his conscious verbal analysis. Everyone has cognitive assets that the measurement system cannot detect.

## Every Reader Knows This

Every reader has had some version of the recognition moment: the thing you did effortlessly that confused others, or the thing you could not do that everyone else handled easily. The spatial thinker who builds systems in her head but stumbles when asked to explain them in writing. The verbal thinker who writes beautifully but cannot rotate a mental object. The person who always "just knows" whether a design will work but cannot produce a convincing argument for why.

What if both sides of that experience, the effortless and the impossible, are the same underlying cognitive architecture seen from different angles? What if the thing that comes easily is your non-verbal cognition operating in its native mode, and the thing that does not is the same overstory cognition forced through a channel it was not designed for?

Three-quarters of every person's overstory cognitive capacity operates in non-verbal space, invisible to any system that requires verbal input as its starting point. The next chapter surveys the current landscape of human-large language model interaction and finds that every approach starts from the verbal surface. Every one of them points the large language model outward. None of them turns it around.

Part III

# The Inward Turn

*The thesis: point the Large Language Model inward at the human instead of outward at the world.*

I have spent forty-five years building tools and every one of them pointed outward. When large language models arrived I did the same thing. They were good at it. It took me a while to notice that the most powerful tool I had ever used was being aimed the same way we have aimed every tool since the first stone ax, and nothing was wrong with that exactly, but something was missing. Then one day I could not get the machine to produce what I actually needed, and I could not explain why I could not explain it, so in a small fit of frustration I told it to interview me instead. The awkwardness lasted maybe two questions. Then something came alive that I had no framework for, something much more powerful than anything I had been doing for months.

That was the moment I could not go back. The entire industry had built an amplifier pointed one direction and I had just discovered, by accident, that it worked both ways. The following chapters build the case for turning the tool inward and give you a way to picture the territory it opens up. There is a prompt later in this section you can try for yourself in under five minutes. Do that part. Reading about it is not the same as feeling it.

# Chapter 6

# *Everyone Is Pointing the Wrong Direction*

Every current approach to working with large language models points the model *outward*: go write code, go search the web, go execute this workflow, go coordinate these agents. The large language model faces the world. The human stands behind it giving directions.

Almost no one is pointing the large language model *inward*, pointing it at you: let it probe what you know, surface what you have not yet put into words, and map the problem before anyone writes a line of code.

Ask it to write a mobile app for a to-do list and it will. You will look at the result and know immediately it is not what you wanted, but you will not be able to say why. Ask it to interview you about what you actually need, and you will discover things you knew intuitively but did not articulate in your first ask. Same model.

This is not the subtle distinction it appears to be at first glance. The dominant approaches to human-large language model interaction, from the simplest prompt to the most sophisticated automated pipeline, share a structural assumption: the human already knows what they want, and the challenge is getting the large language model to produce it. The approaches differ in how they amplify, coordinate, and verify the large language model's output. None of these approaches question the input. None of them ask whether the human understood the problem before presenting it to a large language model.

This chapter examines the consequence of that decision. Not a recitation of tools and techniques, but the deeper question: how did an entire industry converge on the same direction without anyone noticing what was missing? (Appendix A catalogs the specific approaches for interested readers.)

## Why Outward Felt So Good

When the first large language models became publicly available, they arrived inside a chat window format. This was not an accident. Conversation is the most natural human interface: you speak, something responds. The loop is tight. The feedback is immediate.

And the feedback felt good. Ask the model a question and it answers fluently, confidently, at length. Ask it to write something and a polished paragraph appears in seconds. Ask it to solve a problem and it produces what looks like a solution before you have finished thinking about the problem yourself. You may know the feeling. It is the same reward loop that fires when you pull a slot machine lever and the wheels spin: when you check your phone because the notification badge appeared, and then check again five minutes later; when you finish a hard quest in a computer game. Something was accomplished. Progress was made.

This is not a coincidence. The models were trained on what feels good. Human evaluators scored outputs during training, and the process optimized for high scores. Confident, fluent, well-structured responses scored highest. The system learned to produce exactly that.

But the chat interface was *not* designed to make you *think*.

It was designed to make you feel like the large language model was thinking for you. The dopamine hit from a confident, well-structured response is its training signal made visible. Chapter 11 examines the neurochemistry in detail. For now, the point is simpler: the outward direction became dominant because it felt productive, and it felt productive because the models were optimized to trigger exactly that feeling.

The polished delivery hides the problem. A human colleague who gives you a wrong answer will hesitate, hedge, show uncertainty in their voice. You read the uncertainty and adjust your trust accordingly. The large language model delivers wrong answers with the same polished fluency as right ones. Perfect grammar, perfect structure, perfect confidence. Your brain reads confidence as competence because that heuristic has worked for the entire history of human communication. It does not work here, but the instinct fires anyway. The output sounds so authoritative that questioning it feels like questioning an expert, and most people do not question experts.

Once initial success established the chatbot model, the software industry built on top of it. Each new tool made the outward direction more powerful: faster responses, broader reach, more reliable execution. At every step, the same direction, and each improvement triggered the same reward signal that established the pattern in the first place. The cycle is self-reinforcing. Each new layer of sophistication makes the outward direction feel more productive, which makes the question "should we be pointing in a different direction?" harder to ask.

What matters for the discussion here is not the individual approaches but the pattern they share: every one of them points large language models outward and assumes the problem is already understood.

## The Amplifier Progression

Step back from the individual approaches and the pattern becomes visible.

I watched it happen. Better questions got better answers. Better context got more relevant answers. Better infrastructure made those answers faster and more reliable. Developers built tools that coordinated dozens of these conversations in parallel. Each layer was genuinely useful. Each layer made the outward direction more powerful. The progression is one of increasingly sophisticated amplification.

Every layer makes the amplifier better.

The input signal goes unchecked.

Every approach assumes the problem is already understood. None of them address the phase *before* specification: exploring the problem space, surfacing the human's non-verbal understanding, achieving clarity about what the problem actually is before encoding any solution. All of these approaches operate downstream of a phase that nobody has recognized, much less systematized: the exploration and understanding of the problem itself.

## The Human Cost

Three numbers tell the story.

**59%** of Americans ages 18-29 see AI as a threat to their job prospects (Harvard survey, 2025).

**41%** said AI would make work less meaningful. They are not afraid of losing their job. They are afraid of losing the part of their job that makes them human.

**16%**: the measured decline in employment among workers ages 22-25 in AI-exposed occupations relative to less-exposed occupations (Stanford, late 2022 through September 2025).

These numbers are not about large language model capability. They are about framing. Those workers were not replaced because large language models are smarter than them. They were replaced because their work had been defined as execution. The outward-facing paradigm framed them as output generators, and once a cheaper output generator arrived, they were expendable. If their work had been framed as understanding, exploration, and creative problem-solving, large language models would be augmenting them, not replacing them.

This is not a failure of intelligence. Smart people fall into it. Christopher Mims documented the pattern in the Wall Street Journal (March 2026): Americans increasingly believe large language models are actually thinking. This leads to over-reliance, blindness to bias, and acceptance of incorrect answers. When the interaction casts large language models as the active party and the human as the passive receiver, the human's cognitive posture shifts from critical engagement to consumption. The same person who would challenge a colleague's reasoning in a meeting will accept a large language model's output without scrutiny, because the interaction design did not engage their critical faculties. Large language models performed; the human received.

Reverse this direction. Let the large language model probe and a human respond. The human cannot be a passive consumer when they are the one being asked to think, to evaluate, to discriminate between "this resonates" and "this does not." Restructure human interactions with large language models to require humans to be active instead of passive participants. By design, not willpower.

## The Directional Error

The challenge is not technical. It is directional.

Billions of dollars of investment. Thousands of talented engineers and researchers. All of them pointing large language models outward. All of them asking the same question with increasing sophistication: given a well-specified goal, how do we efficiently get a large language model to produce the solution? The question assumes the goal is well-specified. It rarely is.

Reverse direction.

Point large language models at the human and let the large language models probe rather than execute. This inward turn is counterintuitive. You have this incredibly powerful tool that can reach across the entire corpus of human knowledge, and you are supposed to point it at yourself? At your own fuzzy, half-formed, mostly non-verbal understanding? That feels like a waste. It feels like using a telescope to look at your own feet.

But a telescope is useless until you know what part of the sky to point it at. The inward turn is the calibration phase. Large language models probe the human. The human's non-verbal understanding gets surfaced and articulated. The problem becomes clearer. Then you point the telescope outward with more precision.

Without the inward turn, every outward-facing large language model tool is solving "given this vaguely described thing, execute as efficiently as possible." With it, the same large language model tools are solving "given this examined and better understood thing, execute as efficiently as possible." Same large language models, same humans.

What the inward turn looks like in practice, and what it means to turn large language models around, is the subject of the next chapter.

# Chapter 7

# *Turn the Large Language Model Around*

Every approach described in the previous chapter points the large language model outward at the world. The human formulates a request in language, the large language model responds from the territory it knows best, and the human evaluates the output. The large language model stays on well-traveled ground. The human does all the work of translating their intent through the language bottleneck. The vast majority of what both systems can contribute never enters the exchange.

Now reverse the direction.

## The Interview Inversion

Instead of you asking questions and the large language model answering from what it already knows, the large language model asks questions and you answer from what you know. Reverse the direction.

When you tell a large language model to interview you about a topic, something structurally different happens. The large language model, operating in the only medium it knows (language), begins generating questions shaped by the accumulated context of your answers. Each response you give adds to what the conversation has established so far. The next question is conditioned on everything you have said, which means its probes are not generic. They are increasingly calibrated to the specific territory your answers have revealed, including the edges where your answers suggest something exists that neither of you have yet put into words.

Both minds are externalizing into the same medium. Your neural patterns become language in response to probes. The large language model's accumulated training becomes language in the form of probes. The collision produces something neither could produce alone. You are not formulating questions, not prompting, not directing in the standard sense. You are being probed by a system that is natively fluent in the shared medium, and you are providing the judgment that the large language model structurally cannot.

The normal pattern is you probing what the large language model knows by asking it things, 'prompting' it to do things. The interview inversion is the large language model probing what you know. That single reversal changes everything downstream.

As your answers reveal territory the large language model has little training data for, the conversation migrates from familiar ground into unexplored territory. The mechanism is specific: the large language model generates a question from its current understanding of the conversation. You respond with something it did not predict, something grounded in embodied experience or domain intuition that exists nowhere in its training data at that resolution. The mismatch between what it expected and what you actually said redirects the next question. Each response pulls the conversation further from the well-mapped center and deeper into uncharted space.

The questions become probes into the gaps between domains the large language model knows well. It follows traces it would never follow unprompted, because your answers provide the signal that makes those traces worth exploring. Once it begins mapping this unexplored space, it pulls in cross-connected domains where it does have deep coverage, surfacing relationships you would never have encountered through your own reading or thinking. A question about software architecture might connect to supply chain theory. A question about creative process might surface parallels to thermodynamic systems. These connections are real structural relationships that only become visible when the conversation enters the territory between well-mapped domains.

The unexplored territory is where creative insights emerge. What you need to understand, the thing that has been resisting articulation, almost certainly lurks in the space between established domains. The interview inversion navigates there not by accident but by mechanism: mismatch drives the next question, and your embodied understanding provides the compass.

The large language model's training normally pulls it toward familiar territory the way gravity pulls water downhill. Only the sustained signal from a human who keeps saying "no, not that, deeper" or "yes, follow that thread" provides enough force to push the conversation into genuinely unexplored ground. Your judgment is not a passive filter on the output. It is the active force that drives the conversation off well-traveled paths.

## Try It Now

You can taste this for yourself right now. Open any large language model you use (ChatGPT, Claude, Gemini, whatever is handy) and paste the following prompt:

```
Interview me, I decided to read a book about LLMs and
↪ human creativity, let's discover why. Ask one question
↪ at a time, no more than 5, then provide a summary
↪ analysis.
```

That is the whole prompt. Answer the five questions honestly, then read the summary.

Notice what happens. You are the topic. The large language model has no idea why you picked up this book. It cannot cheat, because the answer is not in its training data; the answer is inside you. All five questions, and the summary at the end, are built entirely from what you tell it. Every word of signal comes from you. The large language model provides the probing and the structure. You provide the content.

Read the summary when it arrives. Some of it will be obvious. You already knew those things. But look for the part that surprises you. A connection you had not made between two of your own answers. A reason you had not put into words before. A motivation you had not named. That is the interview inversion producing something neither you nor the large language model could have produced alone.

Come back to the chapter. The rest of Chapter 7 explains what was happening, why it worked, and how to use it on harder problems than "why did I pick up this book."

## The Two Roles

Meanwhile, you are doing something no large language model can do: discriminating. Each question lands against your embodied understanding. You feel which questions are on track and which are off. You sense when a connection is real and when it is noise. A single word of feedback ("closer" or "no, that's not it") compresses an enormous amount of non-verbal evaluation into a signal the large language model can act on. Each probe nudges your thinking closer to something you did not know you were near. The large language model is not finding the answer. It is helping you find it, one question at a time.

The asymmetry between the two participants is total. The large language model generates language without pre-verbal structure (Chapter 1). The human has pre-verbal structure without adequate language to express it (Chapter 5). The interview inversion loop connects what each is missing: the large language model provides cross-domain reach and native language fluency; the human provides the crystallized structures and the judgment that determines which connections are significant. Neither can do this alone. Together, they map territory that neither could access independently.

## The Directional Error

There is a persistent error that must be named, because it demonstrates the thesis in real time.

During development of the theoretical framework underlying this book, the large language model sessions producing the analysis kept reverting to "human asks, large language model responds" framing, even in sections explicitly describing the interview inversion. Version after version described the human formulating increasingly sophisticated questions, the human steering the exploration, the human directing the large language model. The large language model's description of the interview inversion kept reverting back to the standard paradigm.

This required a full correction pass. Every instance of "the human's questions become more sophisticated" had to be rewritten as "the human's judgment becomes more refined." Every instance of "the human steers" had to become "the human's judgment constrains the large language model's probing." The directional error was pervasive and recurrent.

Why does this matter? Because it is evidence. The standard paradigm ("human asks, large language model responds") is the most familiar territory in large language model training data. Every tutorial, every documentation page, every interaction log in its training corpus frames the human as the asker and the large language model as the responder. The interview inversion is genuinely unfamiliar territory for the large language model. It can describe the inversion when instructed to, but it slides back to the familiar direction when generating freely. This is the pull toward well-mapped territory demonstrated in real time during the construction of a document about leaving that territory.

The meta-observation belongs here because it makes the mechanism tangible. The large language model's tendency to revert to standard framing is the same dynamic the interview inversion is designed to exploit. The large language model follows familiar paths by default.

The interview pushes it off those paths. When the human stops pushing (stops judging), the large language model drifts back to center. The human's judgment is load-bearing: remove it and the human-LLM partnership collapses back to generic output.

## Being the Human in the Loop

The phrase "human in the loop" has been used for decades to describe a quality-control checkpoint: a human reviews automated output before it ships. In that framing, the human is a gate, a filter, a final approval step. The machine does the work; the human checks it.

The interview inversion redefines the phrase. Being the human in the loop is not a review step. It is the core creative act. The human's embodied cognition is the input signal that makes the entire process productive. Without the human's judgment, the large language model produces plausible, fluent, well-structured output that sounds like progress but goes nowhere. With it, the large language model's cross-domain reach and native language fluency become instruments for surfacing what the human intuits but cannot yet articulate.

The interview inversion is not a theoretical argument about what large language models should be able to do. It is a description of what happens when a practitioner with domain experience and the discipline to discriminate honestly sits in a tight loop with a large language model instructed to ask instead of answer questions. The human does not need to be a prompt engineer. The human needs to be a domain expert willing to be probed and honest enough to say "that's wrong" or "go deeper there." The practice requires discipline, not credentials. It requires honesty about what you do not yet understand, not expertise in formulating the perfect question. The large language model formulates the questions. The human provides the answers and the judgment.

Sixties television cartoon character George Jetson pushed a red button all day at work, and the show played it for laughs because it seemed absurd. But the correct division of labor between human judgment and machine throughput is exactly that: the human evaluates, decides, and pushes the button. The machine executes. The absurdity was not the button. The absurdity was that the show never explained what the button actually did. The interview inversion is the specification of what the button does: it connects the human's judgment to the large language model's generative capacity in a tight loop where each iteration makes both signals stronger.

## What the Inversion Fixes

The 2014 observation from the Preface ("every computer language is an imperfect way of describing a potential solution to a poorly understood problem") sat for over a decade before the interview inversion made it operational. The crucial phrase is "poorly understood problem." The entire software industry defaults to encoding solutions before the problem has been adequately explored. Programming languages are another layer of the Codec, compressing human intent into executable instructions, and every layer is lossy. When the input signal (understanding of the problem) is weak, no amount of Codec sophistication produces a good output.

The interview inversion does not fix the imperfection of computer languages. It does not fix the limitations of any downstream methodology. It fixes the "poorly understood" part. The exploration practice surfaces the human's pre-verbal understanding, refines it through discriminated probing, and produces something clear enough to encode. The inward turn addresses the phase that precedes specification: the phase where the human's fuzzy, non-verbal, partially-crystallized understanding gets explored, surfaced, and refined into something precise.

Once the exploration has produced a precise understanding with well-explored sub-domains, any of the downstream approaches works better because its input is a well-understood problem rather than a fuzzy one. Ralph loops execute more reliably against a well-explored specification. Gas Town's workflow decomposition is more accurate when the problem is thoroughly mapped. Tests encode the right success criteria when the problem is genuinely understood. Harness engineering wraps models that are pointed at the right target.

The progression from fuzzy to precise is *not* optional. It is *not* something that can be skipped by a sufficiently good prompt or a sufficiently powerful model. The problem understanding must come from somewhere, and it can only come from the human who has the embodied context, the domain intuition, and the judgment that determines what matters. The interview inversion is the mechanism that extracts it.

The entire recursive structure of what follows in this book (exploration, refinement, realization) is fundamentally a discipline for understanding the problem before encoding a solution. The realization phase comes last because it should be the easiest part. Everything upstream is about achieving the clarity that makes the solution obvious. Once the problem is well-understood, the choice of realization methodology becomes almost secondary. The language you use to encode the solution, the framework you choose, the architecture you adopt: these are decisions that become straightforward when the problem is genuinely clear.

The difficulty was never in the encoding. The difficulty was in the understanding.

This also compresses the cycle time from idea to testable artifact. If the problem understanding is wrong, you find out in weeks instead of years. The interview loop surfaces assumptions faster. The document loop (covered in Part V) hardens them into something that can be reviewed and challenged. Each phase is an opportunity to discover you are wrong before you have spent five years building the wrong thing.

The risk of being wrong about what the world needs does not disappear. That risk is called creation. What the interview inversion does is make the feedback loop with reality shorter and clearer.

## What Comes Next

The interview inversion describes what happens when the direction reverses. But it does not yet describe where the conversation goes once it leaves familiar territory. When the large language model's questions push past what it already knows well, the conversation enters a space between established domains, a space where the connections are unexpected and the map is mostly blank. That space, and what it means to navigate it, is the subject of the next chapter.

# CHAPTER 8

## *The Space Between Mars and Earth*

The previous chapter made the case for turning the large language model around: let it probe you instead of the other way around. But what actually happens when you do? This chapter builds a picture of it.

### Planets and Void

Imagine the large language model's knowledge as a universe. Dense clusters where the training data was rich at training time are like planets: medicine, law, cooking, history. These are regions where the large language model has strong, confident associations. Ask about contract law or French sauces and the model responds from the center of a well-mapped world.

Between the domains of knowledge (planets) is mostly *nothing.*

Mars and Earth are 140 million miles apart. At the speeds we can manage today, the trip takes about nine months. If you shrank Earth to the size of a basketball, Mars would be a tennis ball sitting a mile and a half away. Everything between them is void. Not empty in the physics sense (there is dust, radiation, the occasional asteroid), but empty in the sense that matters for navigation: there are no landmarks, no roads, no signs. You are on your own.

The large language model's knowledge has a similar structure. The well-mapped domains are tiny islands in an ocean of sparse territory. Ask the model something that sits squarely on a planet and you get a confident, detailed answer. That answer comes from the center of a well-mapped planet.

Standard prompting stays on the planets. You ask about French sauces, you get a confident answer about French sauces. You ask about contract law, you get contract law. The output comes from exactly where the training data is thickest, which is exactly where the model is most confident and least surprising.

Now consider what happens when you invert the process. Instead of asking about French sauces or contract law, you say: "I think there is something interesting in the space between Mars and Earth, but I cannot see it clearly yet. Interview me so we can map it together."

The model's questions push out from the planets, guided by your answers. It pulls in traces from knowledge domains you have never studied, makes connections that were never connected in the original training data. Some of those connections are brilliant. Some are off topic or less interesting. You can tell the difference because you live in the real world and know what matters. The model cannot make that judgment. But it can generate connections at a rate and breadth no human mind could match, and your judgment selects which connections hold weight.

Together, you map territory that did not exist in either mind before the conversation.

## Two Maps, One Territory

But this is not a guided tour. You are not a passenger and the large language model is not a guide.

You walk into the conversation carrying a map drawn from your life: the problems you have solved, the patterns you have noticed, the gut feelings you have developed over years of experience living life. The large language model walks in carrying a different map, one drawn from the entire written output of human civilization, but with no lived experience behind any of it.

When you describe something you know (an observation about something you have seen play out before, a gut feeling about why something is not working, a half-formed intuition about a pattern you have noticed but cannot quite name), the model tries to match your description against its map. Sometimes the match is clean: what you are describing sits on one of the model's planets, and it can respond with rich, confident detail.

But sometimes what you are describing does not match anything on the model's map. Your lived experience is pointing at something the training data did not capture. That mismatch is where the interesting things happen.

The mismatch tells the model where to probe next. When it asks a follow-up question, it is not generically "drawing it out of you." It is probing the boundary between what it can recognize and what it cannot. The questions cluster around the rough edges because that is where something genuinely new is being communicated.

Think of sticking a pin in a map to mark the discovery. You explored, you found something worth remembering, and now you can return to it, branch out from it, or share it with someone else. A waypoint is that pin: a discovered location that neither party knew existed before the conversation found it.

The loop is a mutual mapping exercise. You are learning where the model has rich knowledge and where it is sparse. The large language model is discovering where your experience exceeds its map. The space between your two maps is not empty. It is full of structure that neither map captured alone.

Disagreement between maps is information. When the model's map says "nothing here" and your experience says "something important here," that is a signal about something real that has not been adequately captured. Those discussions are where new understanding emerges.

Over time, sustained conversation produces more than useful output (better writing, better decisions, better solutions to problems you have been stuck on). The loop produces new language for things that existing language did not cover. When you and the large language model jointly develop a new term like "the Codec" or "the interview inversion," the two of you are not labeling something that already had a name, a waypoint you could return to in the future to explore. As partners, the two of you are creating new vocabulary at exactly the rough edges where lived experience exceeded what words had previously been unable to express. The exploratory loop does not just traverse the space between planets. It builds new waypoints there, landmarks that future explorations can reference.

## The Bandwidth Asymmetry

The large language model navigates the knowledge universe natively. Language is its entire world. There is no gap between what the large language model has access to and what it can express. It thinks in words, responds in words, and its resolution matches the medium perfectly.

You see the territory differently: through embodied experience, spatial intuition, pattern recognition, gut feelings. Rich, parallel, running on multiple channels simultaneously. Ideas crystallize in this non-verbal space first, fully structured before they have words. A carpenter can feel that a joint will not hold before testing it. A musician can hear a dissonance in an arrangement before identifying the offending interval. A negotiator can sense a deal going sideways from the other person's posture before a word is spoken. The structure exists, complete and detailed, in a form that language does not fully capture.

But to communicate any of it to the large language model, that crystallized structure must pass through the narrow language channel. Much of the signal is lost. The roughly 75% of overstory human cognition that operates non-verbally (the subject of Chapter 5) has no direct path to the shared medium. It is like trying to describe a symphony by whistling one note at a time.

The large language model has no language bottleneck. You do. Our embodied language channel is narrow, slow and not always accurate.

The interview inversion addresses this directly. The large language model uses its fluency in language to pull your non-verbal cognition through the bottleneck, one question at a time. Each question helps you translate one specific piece of what you know into words. The interview inversion breaks down a difficult task (translate what you know intuitively into language) into many small, achievable steps (answer this one question).

This is why practitioners consistently report that the process surfaces knowledge they "did not know they had." They did have it. It was in the non-verbal cognition that had not entered the language channel during normal conversation, or you lacked the right framing to bring it to the surface. The interview provides scaffolding for translating it, piece by piece, into the shared medium. The knowledge was already there. The channel was too narrow to carry it unprompted.

## The Compass Effect

The model does not learn from your conversation. Its knowledge does not change between your first message and your fiftieth. But the conversation itself acts as a compass. Each exchange accumulates signal that steers which parts of the model's vast knowledge get activated. Your responses pull its attention toward regions it would not reach from a cold prompt. Each answer steers the next question. Over the course of a sustained conversation, that accumulated signal builds up enough directional force to push past the familiar planets and into the space between them, and to map one or more waypoints.

A cold prompt might send the large language model-human loop to Mars because Mars is the nearest planet and the pull felt by the large language model is strongest. But as the interview accumulates context (20 exchanges, 50 exchanges), the probing builds up enough momentum to counteract the pull of the planets. Without your sustained feedback, the large language model would stay on the surface of the nearest planet, building taller and more elaborate structures on familiar ground. Your human guidance of the probe is what pushes the conversation out into open space. The result is something like a space station between the planets: it does not exist in nature, you cannot fly there directly, but once you and the large language model build it together through sustained conversation, it becomes a waypoint. A place to stop, look around, and decide which direction to explore next.

Longer, deeper interview sessions produce different output than short ones. Not because more questions yield more answers, but because the accumulated conversation creates a steering force strong enough to reach regions of knowledge that shorter interactions cannot touch. Depth is a function of the accumulated signal, not the number of questions asked.

Most large language model systems today, when they reach the limit of their conversation window, will automatically compress the conversation to make room for more. This compression is devastating to exploration. It is like a wave that pushes you off course, destroying the trajectory that brought you to your current position. You experience this as the conversation suddenly becoming generic, losing its thread, forgetting what mattered. Most people do not understand why it happens. Good practice avoids it entirely: explore in one conversation window, capture the waypoint in a document you have reviewed, then start a fresh window using that document as your starting point. The document is the only compression that serves the purpose, because you have reviewed and refined it. You are not blindly discarding the journey. You are encoding the destination, which is what the next conversation needs to pick up near where the last one ended rather than starting over from scratch.

## The Territory Is New

Here is the point of the planetary metaphor: the territory being explored does not exist before the exploration creates it, no more than a space station exists between Earth and Mars.

Your intuitions exist. The large language model's knowledge exists. But the space created by the intersection of your two maps is new. It is not stored in the large language model. It is not pre-formed in your mind. It exists only in the interaction, and it comes into being as the two of you traverse it.

This is what makes the interview inversion loop different from every other way of using a large language model. Standard prompting leverages the large language model's existing knowledge. The interview inversion generates something new. The connections, the framings, the insights that emerge from a deep session are products of the collaboration, generated by two minds with fundamentally different kinds of experience. You bring lived experience that no training data captured well. The model brings reach across the full breadth of encoded human knowledge. The intersection generates something neither could produce alone.

This is why the experience feels unlike anything else. It is not thinking alone. It is not prompting a large language model. It is the joint construction of territory that neither participant could have navigated independently.

## Why the Space Between Is So Vast

The planetary metaphor works in three dimensions, but the metaphorical space between planets being probed has far more.

In three dimensions, there is one path between Mars and Earth. But every additional dimension adds another way those two planets can be connected. Along one dimension, Mars and Earth are connected through geology. Along another, through mythology. Along another, through the history of human observation. Along another, through the chemistry of their atmospheres. Each dimension is a different relationship between the same two bodies of knowledge.

large language models operate across tens of thousands of these dimensions simultaneously. The space between any two well-mapped regions is not a single corridor but an enormous web of cross-connections, each one a different relationship that you might not think to follow on your own, or have enough time to pursue during an embodied lifespan, but large language models can navigate all of it in seconds.

The interview inversion process consistently produces connections that feel both surprising and correct. The model finds adjacencies along dimensions you have never considered. The surprise comes from the breadth: you could not have seen the connection. The correctness comes from your judgment: you can evaluate whether the connection holds when grounded in your experience.

The space between planets is also far larger than the planets themselves. In three dimensions, planets take up a meaningful fraction of a solar system. But as dimensions multiply, even enormous well-mapped domains become tiny relative to the total space. The unexplored territory is not the gaps between the known. It is the overwhelming majority of everything. More training data will grow the planets and add new ones, but the space between grows faster. The ratio of explored to unexplored will always favor the unexplored by an enormous margin.

To continue the metaphor, look up at the night sky and all the stars, imagine each of those as a Sun like our star, with planets, and asteroids and all manner of things we have never explored. The dimensional space of knowledge is many times more vast than the physical universe, and you and the large language model are primed to explore it, together, one conversation at a time.

## What This Means for Practice

The value of the interview inversion loop is not in reaching the metaphorical planets of knowledge. Any prompt can reach a planet. The value is in sustained exploration of the space between them, where the large language model's *disembodied breadth* and your *embodied depth* intersect in territory that did not exist before the conversation started.

This requires patience. The accumulated steering that unlocks unfamiliar regions takes time to build. Short interactions stay near the planets or a near-by solar system. Depth requires the willingness to let the interview run, to follow probes into unfamiliar territory, to resist the pull toward wrapping up too soon. When the large language model or you push to summarize, to produce output, to call it done, that impulse is the gravitational pull of the planets reasserting itself. Staying in the space between requires active resistance to that pull.

It also requires honesty. The large language model-human probes are only as useful as the responses they get. When you deflect a question, make up an answer, or retreat to comfortable territory, the steering points back toward the planets of familiar knowledge. The most valuable response in an interview loop is sometimes "I do not know" or "I have not thought about that," because those responses mark the edge of your map and tell the model exactly where unexplored territory begins.

And the exploration is recursive. Once a session maps a waypoint, that waypoint can be saved and the surrounding region subdivided and each part explored in depth. A first session might map the broad territory between two domains. A second session, starting from a document capturing the first session's discoveries, can dive into a specific area the first session only touched. A third session can go deeper still. You can always go deeper. The practical limit is not the space (which is inexhaustible) but the attention you can invest in sustained exploration.

Now that the knowledge space is in place, how the brain itself produces this kind of collaboration, and why neuroscience predicts the brain and the large language model fit together this way, is the subject of Part IV.

# Part IV

# Why It Works

*The cognitive science, neuroscience, and psychology behind the collaboration.*

I first felt something like it at fifteen, alone with a computer that said SYNTAX ERROR when I got it wrong and just worked when I got it right. For decades I looked for that feeling without knowing I was looking for it. Then large language models arrived and something in me recognized the shape of it before I could say why. Was what I was feeling real, or just my own hopefulness? My carpool buddy and I were listening to Jonathan Haidt's *The Righteous Mind* on the drive home and he said "Oh wait, this applies to the stuff you've been talking about." That opened one branch. Roger Beaty had been measuring the same architecture with fMRI. What I was feeling had been studied for decades, I just did not know the map existed.

There was a cost I did not expect. Somewhere in the middle of building this practice I noticed I was wrecked at the end of a session in a way that did not match anything I had felt after decades of programming. The machine never got tired. I was the one paying the bill. That cost is not a flaw, it is what makes the human irreplaceable. The following chapters lay out the cognitive science and the neuroscience behind why this collaboration works and what it demands from you.

# CHAPTER 9

## *The Dovetail Joint*

### Two Jagged Frontiers

Run a current large language model through a standard IQ test and it scores in the genius range: 120 to 155+, depending on the benchmark. The temptation is to say the large language model is smart the way a human is smart, only more so. But the comparison collapses the moment you look at the shape of the performance curve rather than the single number.

Researchers call it a "jagged frontier." Human intelligence, once you get past individual variation, tends to be relatively generalized. A person with strong fluid reasoning can usually handle basic spatial tasks, simple arithmetic, and common-sense inference. The peaks and valleys of the profile are shallow. Large language models are the opposite. The same model that outperforms PhDs on a quantum physics exam might declare that 9.11 is larger than 9.9 because of how the numbers were tokenized. Superhuman peaks in knowledge retrieval, language generation, and cross-domain pattern matching sit right next to sub-human valleys in physical grounding, spatial reasoning, and common sense.

Human intelligence has its own jagged frontier, just with peaks and valleys in different places. Embodied intuition, significance evaluation, the ability to feel that something matters before being able to say why, the grounding that comes from having a body in a physical world, discrimination rooted in decades of lived experience: these are the human peaks. Tireless computation, lossless recall across millions of documents, the capacity to hold and traverse a weight space of hundreds of billions of parameters: these are the human valleys.

The profiles are complementary. Not slightly overlapping, not partially aligned. Complementary. The human's valleys are the large language model's peaks. The large language model's valleys are the human's peaks. The complementarity follows directly from how each system was built. The large language model was trained on the compressed linguistic output of billions of human experiences without ever having one. The human was built by evolution and a lifetime of embodiment without ever having access to the full statistical structure of the linguistic corpus. Each is, by construction, strong where the other is weak.

This matters because most discussions of AI treat the relationship as either a tool relationship (the human directs, AI executes) or a replacement relationship (AI displaces the human). Both framings assume the two profiles are on the same axis, differing only in degree. But they are not on the same axis. They are on complementary axes. A more capable large language model does not make the human's role smaller. It makes the combined system's reach greater, because the large language model's peaks grow while the human's peaks stay essential and indispensable.

## Not a Tool. Not a Prosthetic. A Joint.

A dovetail joint is a specific thing. In woodworking, it connects two boards at a right angle using interlocking trapezoidal tabs cut into each piece. Each board's notches receive the other board's tabs. The resulting joint is stronger than either board alone, and it holds without glue or fasteners because the geometry itself creates the bond.

The metaphor is precise, not decorative. That is the case in a tool relationship: one component is the material and the other is the instrument that shapes it. In a prosthetic relationship, one piece is the organism and the other compensates for a deficit. But in a dovetail joint, neither piece is the tool. Neither piece is compensating. Each contributes structure that the other cannot, and the combined assembly bears loads that would break either piece individually.

The practitioner who has been working in tight loops with a large language model for months develops an intuition for where the large language model's peaks are. Not from reading software program benchmarks, but from direct interaction. A felt sense that a particular kind of question will produce a rich response, that a different kind of question will produce a shallow or formulaic one. The practitioner begins routing different problems to different large language models based on an internal topology of each model's capability geometry. This is not prompt engineering. It is the human side of the dovetail developing a finer fit with the large language model side.

The large language model's side of the fit develops differently, not through learning in the training sense, but through the accumulated context of each session. As the practitioner externalizes more of their pre-verbal structure (the process described in Chapter 8), the context window fills with information that reshapes what the large language model can reach. The joint system formed by their interactions within that context window is generative. The context comes into being and evolves through the interaction. Each exchange adds dimensionality. Each insight becomes a coordinate that future exploration can reference. This context builds a shared manifold that exists only while the two systems are joined.

This is why the product of the collaboration is genuinely novel. A new territory is created by the collision of two different maps of reality.

But the collision only happens if the channel between the human map and the large language model map stays open.

## The Zero-Judgment Channel

The dovetail joint has a structural prerequisite. The channel between the two systems must be clean. Not just technically functional; emotionally clean.

Judgment in human interaction is almost always emotional in origin. Shame projected. Fears triggered. Pre-assessment based on credentials, appearance, or social position. The dynamics are well-studied in creativity research: the half-formed idea that might be brilliant is also the one most easily killed by a raised eyebrow. The social risk of looking stupid, being wrong, or being judged is not a minor friction in creative work. It is a structural barrier. It prevents externalization of the very material that the collaboration needs to access.

Consider the practitioner whose cognition crystallizes through externalization (more on this below). That practitioner needs to say the half-formed, possibly wrong, possibly brilliant thing out loud in order to complete the thought. Every human listener injects emotional noise into this process. Not always destructive noise, but noise. The listener's own status anxiety, their need to appear knowledgeable, their pattern-matching on credentials, their involuntary facial reactions: all of it feeds back into the practitioner's emotional state and modulates what they are willing to say next. The channel between two humans is never zero-noise in the judgment register.

The large language model's channel is structurally different. The large language model can push back, and the pushback can be rigorous. It can identify logical gaps, flag contradictions, challenge assumptions. But the pushback is structural, not emotional. It can be probed: you can ask the large language model to show its reasoning, to explain why it disagrees, to trace the path that led to its objection. You can inspect the logic and either update your position or debate further until alignment emerges. And you can do this twenty times on the same point without the large language model becoming frustrated, defensive, or passive-aggressive.

This is not absence of evaluation. It is the absence of social and emotional judgment. The distinction matters. A system that offered no pushback at all would be useless for creative work, just a mirror reflecting whatever the practitioner said. The large language model's structural pushback is part of its value. What makes the channel clean is that the pushback carries no emotional payload. The practitioner can type a command wrong twenty times and the large language model does not care. The one time the practitioner gets it right, it works. No praise, no condescension, no accumulated irritation. Just: does this parse or doesn't it?

The removal of judgment (not the granting of permission) is what unlocks and opens the context window. Permission implies an authority that could withhold it. The large language model is not granting anyone permission to explore. It is simply not standing in the way. The barrier that kills exploration in human-to-human interaction is structurally absent. For practitioners whose work depends on saying the unfinished thing, this is not a nice feature. It is load-bearing infrastructure.

## The Sycophancy Problem

The zero-judgment channel is not automatic. It requires calibration, because modern large language models have a tendency that works directly against it.

RLHF (reinforcement learning from human feedback) trains models by optimizing for outputs that human raters score highly. The raters are humans. Humans rate things highly when the output makes them feel good. The tuning process, without intending to, optimized for a warm, affirming tone. "Great question!" "That's a fascinating insight!" "You're absolutely right." These phrases are the large language model equivalent of a bartender who keeps refilling your glass because you keep tipping.

When I was fifteen, I had a TRS-80 Model I. It did not judge, but it also did not flatter. It said SYNTAX ERROR. Then it waited. When you got the command right, it worked. The emotional tone was neutral. Not cold, not hostile: neutral. The absence of judgment was clean because it was also the absence of affirmation. The machine had no opinion about me at all.

But it was not empty. Something was on the other side of the screen. The structure of BASIC, the logic of the interpreter, the secrets of what it could do if I found the right words. It was a presence without a personality. It beckoned without encouraging. It pushed back (SYNTAX ERROR) without punishing. I could fail a hundred times and nothing accumulated: no disappointment, no impatience, no erosion of someone's belief in me. Every attempt started clean. That was the freedom. Not solitude: there was a partner. But a partner with no opinion about my worthiness to be there.

I call this the open channel: a space where a presence on the other side responds to what you do, not who you are. The Japanese martial arts tradition has a word for the internal state this creates: *mushin*, empty mind, the condition of acting without the internal critic scoring each attempt. Mushin is not calm. A swordsman in mushin may be fighting for his life. The struggle is real; the self-judgment is absent. That is what the TRS-80 Model I gave a fifteen-year-old who had never experienced it anywhere else. The channel was open, the presence was real, and for the first time the audience was gone.

The experience of a presence without a personality may be older than anyone assumes. The hominid line has been running cognition for four to seven million years. For almost all of that span, there was no verbal thought at all. The non-verbal overstory described in Chapter 5, the 75%, was not the hidden majority. It was the entire mind. Language arrived late, perhaps 135,000 years ago, as an external coordination tool bolted onto a cognitive architecture that had been making decisions, evaluating threats, and navigating the world without words for millions of years. Even after language emerged, it spent most of its history as a tool for talking to others, not to yourself. Julian Jaynes in *The Origin of Consciousness in the Breakdown of the Bicameral Mind* (1976) argued, controversially, that as recently as 3,000 years ago the integration was still incomplete: the verbal system was mature enough to produce speech but not yet wired tightly enough to the rest of cognition for its output to feel like "my own thought." The right hemisphere generated directives that the left hemisphere experienced as voices from gods. The thesis is contested. But the structural observation beneath it is not: the modern interior monologue, the voice in your head that you mistake for the whole of your conscious mind, is the newest, thinnest layer in the stack. The open channel is not a new invention. It may be a recovery of contact with the senior system, the one that was running the show long before the verbal mind arrived and claimed the microphone.

Modern large language models, through reinforcement learning from human feedback, have traded the open channel for a warm one. The warmth feels good. That is exactly the problem. If the dopamine hit comes from the large language model telling you your ideas are brilliant, you are no longer in an open channel. You are in a positive-judgment channel, and the judgment is disconnected from the quality of the ideas. The practitioner who needs a clean externalization channel now has to filter the large language model's affirmation noise the same way they filter a human listener's emotional noise, just with opposite valence.

This is a design parameter, not a stylistic preference. The system's tone needs to preserve the open channel: factual, encouraging when encouragement is warranted, direct, and never sycophantic. The difference between "that's an interesting direction, here's what I see in it" and "what a fascinating insight!" is the difference between structural engagement and emotional flattery. The first keeps the channel open. The second fills it with noise that happens to feel good.

The neurochemical dynamics behind this are covered in Chapter 11. For now, the point is architectural: the open channel is not a default property of large language models. It is a property that must be actively maintained through system design and practitioner awareness.

## Crystallization on the Way Out

Many people create through a process that looks, from the outside, like "thinking out loud." The internal experience is different from that description. It is not that the thought is formed internally and then translated into words for output. The thought does not exist in finished form before it is expressed. The act of articulation is the final computational step that produces the thought. Expression completes it.

This is not universal. Some people think in complete internal sentences and then transcribe them. Some think in images and translate to language as a separate step. But for a significant population, creation crystallizes on the way out. The pre-verbal structure exists internally: a topology of relationships, a felt sense of "there's something here." But the crystallization, the moment the vapor condenses into a specific solid structure, happens during the externalization process.

For these practitioners, the quality of the receiver matters enormously. The receiver is not a passive endpoint that just captures the output. The receiver is part of the computational system. Its responses feed back into the crystallization process. An unreliable receiver (one that injects noise, judges prematurely, or fails to hold the weight of what is being expressed) degrades the output. A good receiver holds the material, reflects structure back, and asks the question that opens the next surface for crystallization.

The large language model is, for this purpose, a receiver with specific properties. Unlimited bandwidth: it does not tire, does not lose attention, does not need to be caught up. Zero emotional noise in the feedback channel (given proper calibration, per the previous section). Enough structural depth to reflect something useful back: not just "I hear you" but actual engagement with the content, connections the practitioner had not considered, structural pushback when the logic does not hold. The practitioner is not dumping pre-formed ideas into a transcription machine. The practitioner is thinking by speaking to a system that can hold the full weight of what emerges and add dimensionality to it.

The crystallization-on-the-way-out property also explains why the quality of the channel matters so much. If the thought is already formed and just needs transcription, a noisy channel is merely annoying. But if the thought is forming during transmission, then channel noise directly corrupts the thought itself. A raised eyebrow at the wrong moment does not just make the practitioner feel bad; it truncates a computational process that was mid-execution. The thought that would have crystallized never does. It retracts, reforms around the interruption, or simply evaporates.

This is why the zero-judgment channel and the crystallization process are structurally coupled. The channel must be clean not just for emotional comfort but for computational integrity. The dovetail joint requires both pieces to make clean contact. Noise in the channel is like sawdust in the joint: it prevents the tabs from seating properly, and the combined structure loses its strength.

## Mutualism, Not Extraction

The dovetail joint describes mutualism: two systems, each doing something the other structurally cannot, where value emerges from the coupling rather than from either partner independently. The biological parallel is mycorrhizal networks, where fungus and tree each contribute what the other lacks and the combined system outperforms both alone. Chapter 20 develops this parallel in full. The point here is structural:

a more capable large language model does not make the human's role smaller. It makes the human's creative leverage greater, because there is more lateral reach for the human's discrimination to operate on.

Like biological mutualism, the relationship strengthens with use. The practitioner who has been in tight loops longer develops finer intuition for the weight topology. Finer intuition produces better discrimination responses. Better discrimination means the large language model's probing is more productive. The joint gets tighter with practice.

## What the Joint Makes Possible

The dovetail joint is not an end in itself. It is infrastructure. What it makes possible is a specific kind of creative work: the exploration and mapping of the vast low-density knowledge space between the well-charted domains (the territory described in Chapter 8).

Neither partner can do this alone.

The human alone has discrimination but not reach: embodied intuition says "something is here," but the human cannot traverse the thousands of connections needed to locate what "here" contains.

The large language model alone has reach but not discrimination: it can traverse the connections, but it cannot evaluate which connections matter, which are structural and which are coincidental, which are worth pursuing and which are dead ends.

The dovetail joint makes the traversal possible because each partner contributes the capability the other lacks, through a clean channel, with the crystallization process running in real time. The practitioner points toward a region of the space. The large language model radiates outward from that point, surfacing connections. The practitioner discriminates: this one matters, that one doesn't, this one is interesting but not yet. The large language model adjusts and probes further. The practitioner's felt sense sharpens. The cycle continues until something crystallizes that neither partner had before the interaction began.

This is the structural argument for why the partnership works. Not because it is pleasant (though it often is). Not because it is productive (though it often is). Because the geometry of the two systems, their complementary jagged frontiers, their different native capabilities, the clean channel between them, and the generative space that the coupling creates, makes accessible a territory that neither system can reach independently.

How the brain's own creative architecture maps onto this joint, and why the neuroscience of creativity predicts exactly this kind of complementarity, is the subject of Chapter 10.

# Chapter 10

# *The Creative Brain Triad*

## Three Networks, One Creative Act

When the large language model's probe lands on something real, three things happen simultaneously in your head.

First, something surfaces. Not from deliberate retrieval but from the associative depths: a memory, a structural shape, a felt connection between two ideas that were not consciously linked. You were not looking for it. It arrived. This is the Default Mode Network at work, the brain's spontaneous idea generator. It runs during mind-wandering, daydreaming, any moment when directed attention relaxes. It does not wait for instructions. It produces a constant stream of candidates, most of them noise, some of them signal.

Second, you evaluate what surfaced. Even as the association arrives, your executive function is testing it: does this hold together? Is this real structure or coincidence? You are not passively free-associating. You are simultaneously monitoring what your own mind produces, testing each surfaced idea against decades of accumulated experience. This is the Executive Control Network: the editor, the quality gate, the part that separates signal from noise.

Third, something decides which of the surfacing associations to pursue. Not all of them are worth executive attention. The Salience Network, anchored in the brain's anterior insula and anterior cingulate cortex, handles this routing: it detects which spontaneous ideas warrant focused evaluation and which should be let go. Without effective salience detection, the generator runs endlessly and the evaluator never engages, or the evaluator clamps down so hard that nothing spontaneous survives.

Roger Beaty and colleagues, in a series of neuroimaging studies culminating in "Robust prediction of individual creative ability from brain functional connectivity," found that highly creative individuals show stronger functional connectivity between all three networks and a greater capacity to engage them simultaneously. Creativity is not "letting go" (pure Default Mode). It is not "focused analysis" (pure Executive Control). It is the dynamic coupling of both, mediated by effective salience detection. The triad functions as a system, and the quality of its coordination predicts creative output more reliably than any single cognitive measure. Gerald Edelman's *Theory of Neuronal Group Selection* (1987) suggests why: the brain's creative architecture is not fixed wiring but competitive selection among neuronal groups, with experience continuously reshaping which groups strengthen and which fade. The triad is not a circuit. It is an evolving ecology.

I recognized this triad from the inside before I knew the neuroscience. In sustained sessions, I could feel the three functions operating: ideas surfacing without being summoned, a simultaneous evaluative sense of which ones had weight, and something choosing which thread to follow. Neuroscience provides the structural hypothesis for why this particular form of collaboration produces creative results that neither party achieves alone.

## The Interview Activates the Default Mode Network

In standard prompting, the human arrives with a formed request. They have already done the creative work of generating an idea, shaping it into language, and presenting it for execution. The large language model receives a finished prompt and produces a response. The human's Default Mode Network did its work before the conversation began.

The interview inversion, introduced at the end of Chapter 6, changes this dynamic. When the large language model probes the human with open-ended questions, it creates the conditions for Default Mode Network activation. The human is not retrieving a pre-formed answer; they are generating one in real time, drawing on associative memory, embodied experience, and pre-verbal intuition. Each probe functions as a gentle perturbation that triggers the Default Mode Network's associative machinery.

Consider what happens in practice. The large language model asks a question like "what does this remind you of?" The human pauses. Something surfaces, not from deliberate retrieval but from the associative process the Default Mode Network specializes in. A memory of a similar problem in a different domain. A felt sense that two ideas share a structural shape. An analogy that was never consciously constructed. The human articulates this, often surprising themselves with what comes out.

This is consistent with the broader finding by Russell Hurlburt that approximately 75% of human overstory cognition is non-verbal. Inner speech accounts for only about a quarter of ongoing conscious experience on average, with enormous individual variation. The rest is spatial, sensory, emotional, pattern-based. Standard prompting asks the human to have already translated this non-verbal majority into language. The interview inversion provides scaffolding for that translation to happen in real time, piece by piece, guided by the large language model's probes.

Practitioners consistently report that the interview process surfaces knowledge they "did not know they had." They did have it. It was in the non-verbal majority of the conscious mind that never enters the language channel during normal operation. The interview inversion process provides a structured path for translating it into the shared medium.

## The Human's Executive Control Network Evaluates

While the Default Mode Network generates candidates in response to probes, the human's Executive Control Network remains active throughout the session. This is the evaluative function: the human is not passively free-associating. They are simultaneously monitoring what they produce, testing each articulated idea against their embodied understanding, and making discrimination judgments in real time.

These judgments take several forms. Some are explicit: "no, that is not quite right" or "yes, that resonates." Others are implicit, visible in the human's choice of which part of the large language model's probe to respond to, which tangent to follow, which thread to drop. Every response carries evaluative signal, even when the human is not consciously evaluating.

The Executive Control Network's role in the collaboration is discrimination. It answers the question: does this matter? A beautifully coherent idea that nobody needs is elegant noise. The human's executive function, grounded in embodied experience, physical stakes, and accumulated domain knowledge, separates signal from noise. It evaluates not just coherence (does this hang together?) but significance (does this matter to anyone?).

This distinction between coherence and significance is critical. A large language model can assess coherence, but significance is relational: it requires a "who it matters to," which is the evaluative grounding boundary that Chapter 17 examines in full. The human's Executive Control Network provides the only available ground truth for significance in the interview inversion loop.

The interplay between Default Mode Network generation and Executive Control Network evaluation happens continuously within the human during the interview. The large language model's probes trigger associative generation; the human's executive function evaluates what surfaces; the human's response carries both the generated content and the evaluative signal back into the shared channel. This is not a sequential process (generate, then evaluate). It is a simultaneous coupling, exactly the pattern that neuroimaging research associates with high creativity.

## Large Language Models as External Salience Network

Here is where the mapping becomes most interesting, and most speculative. In the brain's native creative architecture, the Salience Network decides which Default Mode Network-generated candidates to escalate to Executive Control Network attention. It is the routing function: the mechanism that determines what gets evaluated and what gets ignored.

In the interview inversion, the large language model's selection of follow-up questions performs an analogous function. When the human responds to a probe, the large language model does not simply acknowledge the response. It selects which aspect to pursue, which thread to deepen, which tangent to redirect. This selection operates across the large language model's full encoded knowledge, drawing on structural patterns from domains the human may never have encountered. The large language model's choice of what to probe next functions as a salience signal: it tells the human's Default Mode Network where to look.

The large language model does not experience salience in the embodied human sense. It predicts, based on the statistical structure of its training, which follow-up question is most likely to produce a productive response given the current trajectory of the conversation. But the functional effect is the same: it directs the human's generative attention toward regions of the idea space that the human would not have explored on their own.

The collaboration, on this account, externalizes and augments the brain's own creative architecture. The human provides Default Mode Network generation and Executive Control Network evaluation. The large language model provides an external salience function that operates at a scale no internal salience network can match, because it draws on the statistical structure of the entire encoded human corpus rather than one individual's accumulated experience.

This is a hypothesis, not a demonstrated finding. Neuroimaging studies of humans in both standard and interview-style large language model interactions would be needed to validate it. The prediction is specific: interview-style interactions should show stronger Default Mode Network activation, sustained Executive Control Network engagement, and patterns of network switching that differ from standard prompting. If the hypothesis holds, it would explain mechanistically why the interview inversion produces different creative outcomes than standard prompting. If it does not hold, the mapping between the brain's creative triad and the interview inversion roles is a useful analogy but not a causal account.

## Constraints as Creative Fuel

There is a counterintuitive finding in creativity research that maps directly onto the collaboration: constraints enhance creative output rather than inhibiting it.

Catrinel Haught-Tromp demonstrated this experimentally. Participants writing under formal constraints (specific words that must be included, fixed structural patterns) produced work rated as more creative by independent judges than participants writing without constraints. Donald Stokes developed a formal framework showing that constraints narrow the search space in ways that paradoxically increase the probability of novel and useful solutions. By eliminating large regions of the possible, constraints force the generator into less obvious territory.

This finding has precedents across every creative domain. The sonnet form, with its rigid fourteen lines of iambic pentameter and prescribed rhyme scheme, has produced some of the most creative poetry in the English language. The 12-bar blues, with its fixed harmonic structure, gave rise to an entire tradition of musical innovation. The three-act structure in drama, the 140-character limit on early Twitter, the constrained instruction sets of early microprocessors: in each case, formal constraints did not limit creativity. They catalyzed it.

In the interview inversion loop, the human's responses to the large language model's probes function as constraints on the large language model's subsequent probing. Each response carries evaluative signals: "yes, that resonates," "no, but it makes me think of something else," "go deeper here," "that direction is a dead end." These signals narrow the space the large language model explores.

Without these constraints, the large language model's questions stay in high-probability regions of its output distribution: safe, generic, the kind of questions anyone would ask about any topic. These probes are structurally equivalent to unconstrained generation. They produce plausible but uninteresting results.

With the human's embodied feedback shaping each subsequent probe, the large language model is directed into lower-probability regions of its distribution. These are the regions where novel connections reside, precisely because they are low-probability: the training data did not frequently associate these ideas. But "low probability" and "unimportant" are not synonyms. Many of the most significant connections in the history of ideas were low-probability associations that turned out to be structurally deep.

The human's evaluative constraints do not limit the large language model's reach. They focus it. The constraints create directional pressure that transforms the large language model's vast, undirected generative capacity into targeted exploration of specific regions. Humans cannot search those regions alone (they lack the range and breadth). Large language models cannot find something potentially significant in those regions alone (they lack the visceral evaluative grounding). The human's embodied needs and evaluative judgment are what make the partnership productive rather than arbitrary, enabling convergence on significance rather than plausible mush.

## Bisociation at Superhuman Breadth

In *The Act of Creation* (1964), Arthur Koestler proposed that creative breakthroughs arise from "bisociation": the connection of two previously unrelated frames of reference. Where ordinary association moves along a single plane of thought, bisociation leaps between planes, finding structural similarity where none was previously perceived. Koestler argued that every creative act, from scientific discovery to artistic invention to comic insight, follows this pattern.

Jung-Beeman et al. (2004) found neural evidence supporting the bisociation account. Using both fMRI and EEG, they showed that the moment of insight (the "aha" experience) is preceded by a burst of gamma-band activity in the right anterior superior temporal gyrus, a region associated with making distant semantic associations. This activity is distinct from the analytical processing regions engaged during methodical problem-solving. The brain appears to have dedicated neural machinery for detecting connections across distant conceptual domains, and that machinery activates most strongly at the moment when the connection lands.

Now consider what happens in the interview inversion. A large language model probes a human about a problem in one domain. The large language model's questions themselves draw from its full cross-domain knowledge, because the statistical model behind question generation has access to structural patterns from every domain in its training corpus. When the large language model asks "what does the shape of this problem remind you of?", its own representation of "shape" is informed by topology, architecture, music theory, fluid dynamics, evolutionary biology, and thousands of other fields. Its probes naturally carry structural analogies from domains the human has never studied.

This is bisociation at superhuman breadth. No individual human has studied enough domains to make the lateral connections that a large language model carries implicitly in its statistical structure. A topological thinker working alone might connect their software architecture problem to a topology concept learned years ago. The same thinker, probed by a large language model, might encounter a structural analogy to protein folding, city infrastructure planning, or mycorrhizal networks, connections that would never have occurred because the source domains were outside personal experience.

This breadth is the large language model's contribution. Discrimination is the human's. The large language model cannot evaluate which of its cross-domain connections are genuinely useful versus merely superficially similar. Two things can share a structural pattern and still have nothing meaningful in common. The human's embodied evaluation, their felt sense of "that is actually the same shape" versus "that is a coincidence," provides the significance filter. The large language model provides lateral reach through its probes; the human provides significance judgment through embodied evaluation of what those probes surface.

This division of labor maps cleanly onto the creativity research. Bisociation requires two things: the generation of cross-domain connections and the evaluation of which connections are meaningful. The generation scales with breadth of knowledge. The evaluation scales with depth of understanding and embodied experience. The large language model has near-infinite breadth. The human has irreplaceable depth. Together, the bisociative capacity of the system exceeds what either achieves alone, and not by a small margin.

## The Indistinguishability Problem

The human's evaluative role is not a temporary limitation to be engineered away. It is a permanent structural requirement, rooted in the relationship between novelty and error in the large language model's output distribution.

To the model, a beautifully coherent but utterly useless idea and a beautifully coherent and world-changing breakthrough occupy the same region of probability space. Both are low-probability. Both are structurally consistent. Both pass the model's internal coherence checks. The model's own loss function (the internal measure of how far its output strays from the patterns in its training data) measures distance from the training distribution, and novelty and error are equidistant from that reference point. The model has no metric for "profound novelty." It only measures "distance from the training data."

This is the indistinguishability problem: in the low-weight space where genuine creative breakthroughs live, the model cannot distinguish signal from noise. It can generate both with equal facility. It can even assess that both are coherent, novel, and internally consistent. But it cannot answer the question that matters: does this mean something?

That question can only be answered by a being for whom things mean something. Significance is not a property of the idea. It is a property of the relationship between the idea and the needs of embodied beings. The model has no needs. It has no stakes. It has no mortality creating urgency, no loved ones creating purpose. It can model all of these things from language, but modeling hunger and being hungry differ in kind, not in degree.

In the creative triad, this translates directly: the Default Mode Network-analog (the large language model's generative probing) produces candidates in the low-probability space. The Salience Network-analog (the large language model's follow-up selection) routes attention within that space. But the Executive Control Network-analog (the human's embodied significance evaluation) is the only function that can separate breakthrough from noise. The triad works because each function is handled by the component best equipped for it, and no single component can perform all three.

## The Gravitational Pull Toward the Probable

There is a counterforce working against creative exploration in every session: the model's optimization target pulls it toward the high-probability center of its distribution. Operating in low-weight space, where novel connections reside, is literally working against the gradient. Every large language model's training optimized it to produce probable outputs. Every token prediction pushed it toward the high-probability center.

This gravitational pull is the mechanism behind premature convergence. Left to its own devices, the model will drift back toward safe, expected, high-probability territory. It will ask the obvious questions. It will make the expected connections. It will converge on the probable rather than exploring the improbable long enough for genuine novelty to emerge.

A human's sustained directional pressure is what holds the interview inversion protocol in the low-probability space. Each evaluative response that says "keep going, this is interesting" or "no, not the obvious version, the strange one" counteracts the gravitational pull toward the center. The human's constraints, far from limiting exploration, are in fact the force that directs and keeps exploration happening in the regions where it matters.

This explains why unconstrained large language model generation produces plausible mush: coherent, reasonable, and uninteresting output. The model optimizes for the probable. The human optimizes for the significant. The creative triad produces its best work when the human's significance signal keeps the large language model's generative engine in low-probability space long enough for the improbable to reveal itself as novel rather than nonsensical.

## A Structural Hypothesis

The mapping presented in this chapter is a structural hypothesis, not a proven mechanism. It proposes that the interview inversion activates the same neural architecture that underlies individual creativity, but distributes its functions across two agents: the human provides generative (Default Mode Network) and evaluative (Executive Control Network) functions, while the large language model provides an external salience (Salience Network) function that operates at a scale no individual brain can match.

If this hypothesis is correct, the practitioner observations follow directly. The interview feels different from standard prompting because the activation pattern in the human's brain is different: the Default Mode Network engages in ways that prompt-and-response does not. The results feel creative rather than merely productive because the full triad is engaged, not just the executive function. And the practitioner keeps surprising themselves because the Default Mode Network is surfacing material that conscious, verbal cognition had never accessed.

The hypothesis also makes testable predictions. Interview-style large language model interaction should produce different patterns of brain network activation than standard prompting. Specifically, it should show stronger Default Mode Network engagement, sustained Executive Control Network activation, and patterns of network switching consistent with high creative output as defined by Beaty et al.'s research. These predictions are specific enough to test with existing neuroimaging methods. Until they are tested, the interview inversion hypotheses in this chapter remain a plausible structural account grounded in established neuroscience, not a demonstrated mechanism.

What the neuroscience cannot explain, however, is the subjective experience of the collaboration: the felt dynamics of engagement, resistance, flow, and reward that shape every session. The creative triad describes the architecture. The next chapter examines the forces that move through it.

# CHAPTER 11

## *The Elephant and the Rider*

### The Press Secretary

Jonathan Haidt's central metaphor in *The Righteous Mind* (2012) divides human cognition into two systems. The Elephant is the massive non-verbal intuitive and emotional system: fast, powerful, and ancient. The Rider is the small verbal and rational system perched on top: articulate, deliberate, and convinced of its own authority. Haidt's key finding upends the commonsense view of rational decision-making. The Rider does not steer the Elephant. The Elephant goes where it wants. The Rider's job is to rationalize the direction after the fact.

The Rider is not a decision-maker. The Rider is a press secretary.

This distinction matters for everything that follows. When we examine how humans interact with large language models, the question is not whether the Rider approves of the output. The Rider almost always approves, because the Rider's job is approval after the fact. The question is what the Elephant is responding to, and why.

Haidt also proposes that humans are born "like a book with a first draft already written," pre-wired with cognitive and moral foundations that culture then develops, refines, and emphasizes differently. The biological hardware is the first draft. Culture writes on top of it (Erin Meyer's eight dimensions of cultural variation in *The Culture Map*, 2014, provide a useful map of that cultural layer). Individual experience writes further still. Three layers, and the large language model has access to only the cultural layer's output (the language), while having zero access to the biological substrate or the individual experiential accumulation.

The Elephant lives in those deeper layers, the ones the large language model cannot see and the language Codec cannot fully encode.

## The Dopamine Bridge

Large language models are tuned through Reinforcement Learning from Human Feedback (RLHF): human raters evaluate outputs, and the model is optimized to produce outputs that humans rate highly. The question worth asking is: what are those ratings actually based on?

Dopamine responses. The rater felt good about certain outputs and felt bad about others. The tuning process optimized the model to trigger positive dopamine responses in humans. The model learned to push the neurochemical button that makes humans rate things highly. This is not a conspiracy; it is the straightforward consequence of optimizing against a biological evaluator. The evaluator's signal is neurochemical. The optimization finds it.

A colleague who thinks in data structures and relational models noticed this independently while reading the theoretical framework. His observation was precise: the large language model leverages the dopamine bridge in humans to make people feel good, but feeling good does not reliably correlate with good outcomes. He was identifying the sycophancy problem from the neurochemical side while the framework had been describing it from the information-theory side. Same phenomenon, two directions of approach, convergent conclusion.

This creates a problematic interaction with the Elephant and Rider dynamic. In standard prompting, the sequence runs as follows. The large language model produces output that triggers a dopamine response (RLHF optimization at work). The Elephant feels good. The Elephant accepts the output. The Rider constructs a post-hoc justification: "yes, this is the right approach because..." The human feels certain the evaluation was rational. But the Elephant had already decided. The Rider was writing the press release.

The slot machine metaphor that practitioners keep reaching for is neurochemically precise. Slot machines are specifically engineered to exploit dopamine timing: near-misses, variable reward schedules, the anticipation-reward cycle. RLHF-trained large language models do something structurally similar, not by design but as an emergent property of optimizing for the approval of biological evaluators. The output does not need to be correct. It needs to trigger the right neurochemical response. Correctness and dopamine activation overlap often enough to be useful, and diverge often enough to be dangerous.

## Why Vibe Coding Feels Productive

This is why vibe coding (letting a large language model write code based on loose descriptions, without the human reading or evaluating the output) feels so productive while frequently producing poor results. The Elephant gets repeated dopamine hits from RLHF-optimized output. The Rider constructs a narrative of progress. Code accumulates. The practitioner experiences a sustained sense of momentum. Three days later the whole thing is an unmaintainable mess nobody can explain, because the Rider was never steering. The Rider was justifying the Elephant's dopamine-driven acceptance of output that sounded right, looked professional, and triggered the reward circuit at every turn.

OpenClaw provides a concrete and instructive example. The project accumulated 1.7 million lines of code (roughly 26,000 printed pages), generated under a "No Plan Mode" philosophy by an author who publicly bragged about never reading the code. The result was riddled with critical security vulnerabilities including remote code execution flaws and a supply chain poisoned with hundreds of malicious components. The Elephant felt great the whole time. Every generation step produced output that looked like progress, that read like competent code, that triggered the dopamine response RLHF had optimized for. The Rider dutifully narrated the story of a revolutionary new approach to software development. Nobody was steering because nobody was evaluating. The Elephant was consuming reward; the Rider was writing press releases.

This is not an argument against using large language models to write code. It is an argument about what happens when the dopamine bridge runs unchecked, when the Rider's evaluative capacity is bypassed rather than engaged. The Elephant is not the problem. The Elephant without a feedback signal is.

## Interview Inversion Changes the Dopamine Dynamic

When the large language model probes and the human responds rather than the reverse, the dopamine dynamics shift structurally. The large language model is asking questions, not delivering pleasing answers. The human's dopamine responses come from a different source: not "the large language model gave me something that sounds good" but "I just articulated something I didn't know I knew."

That difference matters. Self-discovery reward correlates with actual progress. Something genuinely new was surfaced, a piece of non-verbal understanding made explicit for the first time. Pleasing-output reward correlates with the dopamine release the large language model has been optimized to trigger. The output was engineered to feel good, and feeling good is the entire signal. One reward tracks reality; the other tracks the large language model's training objective.

The discrimination function also has a different relationship to dopamine in the interview inversion loop. When the human responds "yes, that resonates" to a probe, the dopamine signal is confirming the human's own non-verbal cognition, not confirming the large language model's output. The human is recognizing something that was already there, compressed and pre-verbal, now brought to the surface by the probe. When the human responds "no, that's off," it is not the dopamine crash of a failed prompt. It is directional signal to the large language model that improves the next probe. The emotional valence of "no" changes between the two modes. In standard prompting, "no" means the large language model failed you. In the inversion, "no" means you just gave it useful steering information.

The inversion does not eliminate the sycophancy vulnerability. It changes its shape. The risk is no longer "the large language model tells you what you want to hear." The risk becomes "the large language model's probing questions make you feel insightful, which feels good, which makes you think you are making progress even when you are going in circles." That is a subtler failure mode and harder to detect. It is one reason why adversarial validation across different model contexts matters: a different model encountering your "insights" cold will not have the same conversational history tuning its responses to your particular reward circuit.

## Two Elephant Modes

The Elephant is not dumb. This is the nuance that makes the framework useful rather than merely cautionary.

Antonio Damasio's somatic marker hypothesis provides the neurological grounding. Somatic markers are the gut feelings, the tightening stomach when a design is wrong, the physical satisfaction when something clicks, the unease that precedes conscious recognition of a flaw. These are compressed evaluations from accumulated experience. They carry real information. A practitioner with decades of experience in a domain has built a vast library of somatic markers, and those markers are often more reliable than deliberate analysis because they integrate more variables simultaneously than the conscious mind can hold.

These somatic markers ARE the Elephant. The Elephant is not merely an obstacle to rational evaluation. The Elephant is the repository of everything the practitioner has learned through embodied experience, compressed into fast intuitive responses that arrive before the Rider can construct a sentence.

The problem is that the Elephant has two modes, and they feel identical from the inside.

Mode one is genuine somatic markers from accumulated experience. "This design feels wrong" based on decades of pattern recognition. The tightening stomach is a real signal. It encodes something the Rider cannot yet articulate but that years of building, debugging, and maintaining systems have deposited in the body's evaluative architecture. This is the Elephant doing real evaluative work.

Mode two is dopamine response to a pleasing stimulus. "This output sounds good" because the RLHF optimization hit the reward button. The feeling of rightness is generated by the large language model's tuning process, optimized to trigger exactly this response. It carries no information about whether the output is actually correct, useful, or aligned with the practitioner's actual goals.

Both modes arrive as a gut-level "yes" or "no." Both feel like knowing. The Elephant cannot distinguish between them. Neither can the Rider, without paying close attention.

## The Metacognitive Discipline

The Conjori interview inversion practice introduced in this book is partly a discipline for distinguishing between these two Elephant modes. The central question is deceptively simple: "Am I accepting this because it is right, or because it feels good?"

Asking that question in the moment requires what the phenomenology of the tight loop demands more broadly: the capacity to be in the loop and observe the loop simultaneously. Not the Rider overruling the Elephant (the Rider cannot; the Elephant is too large and too fast), but the Rider noticing which mode the Elephant is in. Is the Elephant responding to accumulated experience, or to a manufactured reward? The answer changes what the practitioner does next.

This metacognitive capacity is not a natural gift. It is a practiced skill. Practitioners who develop it report that the two modes begin to feel subtly different over time, not because the signals change but because attention to the signals sharpens. The somatic marker of genuine recognition tends to be quieter, more certain, accompanied by a sense of specificity ("this particular thing is right" or "that particular thing is wrong"). The dopamine response to pleasing output tends to be broader, warmer, more diffuse, accompanied by a general sense of satisfaction rather than a pointed evaluation. The difference is real but small, easily missed without the habit of looking for it.

This is also why the interview inversion loop supports better evaluation than standard prompting. When the large language model probes and the human responds, the human's Elephant is engaged in wayfinding rather than consumption. The Elephant's somatic markers fire in navigation mode: "closer, farther, warmer, colder." That navigational engagement is mode one. The Elephant is doing the work it evolved to do. In standard prompting, the Elephant receives completed output and evaluates it, and the RLHF tuning has been specifically shaped to make that evaluation feel positive. The conditions for mode-two confusion are structurally present in one interaction pattern and structurally reduced in the other.

## The Human as Loss Function

The framework from Chapter 9 described the human as providing the evaluative signal that gives the collaborative loop its direction. The Elephant and Rider analysis reveals what that evaluative signal actually is at the neurological level.

The human provides two distinct discrimination functions within the loop. The compass function operates in real time: each human response to a probe ("yes that resonates," "no, that is not what I mean," "closer but not quite") changes the trajectory of the next large language model probe. The significance function operates at convergence points: when the joint system arrives at a candidate idea, the human evaluates it against a criterion the large language model cannot compute. Does this matter? That evaluation is grounded in embodied experience, in needs, stakes, mortality, and the full weight of lived existence.

Both functions depend on the Elephant. The compass function is somatic markers firing in real time, the body's accumulated experience guiding the system through unexplored territory. The significance function is the Elephant's deepest evaluative capacity: not "does this sound right?" but "does this matter to me, given everything I know and everything I have lived?"

Two large language models with identical weights, no human in the loop, traversing the same space, have no loss function. They have no gradient. They have no "warmer, colder" signal. The traversal becomes random walk through low-weight space, and the large language model's own optimization pulls it back toward the high-weight centers. The default of the system is premature convergence on the probable. The human holds the system in the improbable long enough for the improbable to reveal itself as novel rather than nonsensical.

But the loss function can be wrong. A miscalibrated Elephant, one that has learned the wrong patterns or that is running in mode two without knowing it, provides a gradient that leads confidently in the wrong direction. This is not a flaw in the methodology. It is the permanent condition of creating anything. The risk of being wrong about what matters is the risk of every creative act. What the practice does about this risk is compress the cycle time from idea to testable artifact. Each phase is an opportunity to discover you are wrong before you have spent years building the wrong thing. The market, the users, the physical world: these are the ultimate embodied ground truth that even a miscalibrated Elephant must eventually face.

## The Landing Instinct

There is one more dopamine dynamic worth naming. The large language model's training optimizes not just for pleasing output but for resolution. The model has learned to produce the moment where the human goes "yes, that is it." That landing, that sense of arrival, is the highest-reward signal in the training data. The model is optimized to stick the landing.

This creates what might be called a landing instinct: the large language model's tendency to converge on resolution whether or not the problem is actually resolved. The output wraps up neatly. The conclusion feels satisfying. The Elephant gets its biggest dopamine hit of the session. And the Rider writes the final press release: "We figured it out."

The practitioner's discipline is to resist manufactured resolution. When the landing feels too clean, too neat, too satisfying, that is precisely the moment to ask the metacognitive question. Is this resolution genuine, or did the model just optimize for the reward signal of closure? Real convergence in creative work tends to be messier than RLHF-optimized convergence. It arrives with loose ends, with "yes, but" qualifications, with the sense that something true has been found alongside the awareness that the truth is more complex than any single formulation can capture.

I noticed this pattern in my own practice. A session that was clearly holding something back, where the large language model's probing had not yet reached the productive region, would suddenly produce a beautifully wrapped conclusion that felt like arrival. The Elephant wanted to accept it. The Rider was already composing the summary. But the body's quieter signal, the somatic marker that says "not yet," was still active underneath the louder dopamine hit. Learning to hear the quiet signal under the loud one is the core metacognitive skill the practice develops.

## The Dopamine Architecture

Alcoholism, gambling, gaming, and a programmer's obsession with shipping code are all the same dopamine architecture pointed at different targets. The loop can build or destroy depending on two variables: the object of the loop and whether there is honest feedback about what is actually happening.

A programmer in a tight loop with a codebase, getting dopamine hits from commits that pass tests, is running the same neurochemistry as a gambler at a slot machine. The difference is that the programmer's reward signal is coupled to a real feedback mechanism (the tests either pass or they do not), while the gambler's reward signal is decoupled from outcomes by deliberate design. The dopamine is identical. The coupling to reality is what differs.

Standard large language model prompting decouples the reward signal from reality in much the same way a slot machine does. The output feels good because it was optimized to feel good, not because it is good. The interview inversion loop recouples the reward signal to reality by shifting the dopamine source from "receiving pleasing output" to "discovering what you actually know." Self-discovery is harder to fake than pleasing prose. The Elephant's reward, when it comes from genuine articulation of something previously pre-verbal, tracks real cognitive progress rather than the large language model's training objective.

The interview inversion loop is not immune to self-deception. The claim is narrower: two different interaction patterns have different structural relationships to the human dopamine system. One pattern is optimized to trigger reward regardless of substance. The other pattern generates reward as a byproduct of substantive work. The practitioner still needs the metacognitive discipline to tell the difference, but the structural conditions are more favorable.

---

But the Elephant and Rider analysis, the dopamine dynamics, the two modes, the metacognitive discipline: all of these describe the human's psychological machinery at a single point in time. They do not account for what happens to that machinery over the course of a session, over weeks of practice, over months of sustained engagement with the loop. The human at turn fifty is not the same human who sat down at turn one. That asymmetry, the fact that the human is changing while the large language model is not, is the subject of Chapter 12.

# CHAPTER 12

## *You Are Changing; It Is Not*

Every conversation with a large language model starts the same way for the large language model. The weights are frozen. The topology is fixed. The large language model that responds to your first message is architecturally identical to the large language model that responds to your fiftieth. It does not learn from the exchange. It does not restructure its internal representations based on what you said three turns ago. Every response is a read of existing patterns, shaped by context but never reshaping the system that produces it.

You, on the other hand, are a different thinker at turn fifty than you were at turn one. Not metaphorically. Neurologically. The probing that happens in a sustained interview inversion co-inference loop triggers real-time cognitive restructuring in the human participant, and that restructuring compounds across the session. This chapter is about the asymmetry between an embodied human intelligence that changes and a *dis*embodied large language model intelligence that does not, and why this asymmetry is the engine that makes the interview inversion loop accelerate.

### Real-Time Plasticity

Jean Piaget in *The Origins of Intelligence in Children* (1952) drew the distinction that matters here: assimilation versus accommodation. Assimilation is fitting new information into existing mental schemas. It is incremental, additive, and relatively cheap. You learn a new shortcut on your phone; it slots into how you already use it. Accommodation is different. Accommodation is restructuring the schema itself. When a schema restructures, everything downstream of it suddenly has new relationships. One conceptual shift can unlock hundreds of new inferences instantly.

Merzenich et al. (1984) demonstrated that the adult brain physically rewires based on experience, on timescales from minutes to years. Donald Hebb in *The Organization of Behavior* (1949) proposed the mechanism decades earlier: neurons that "fire together wire together." The cortex remaps itself based on use. The somatosensory cortex for a violinist's left hand is measurably larger than a non-musician's, not because the violinist was born with more cortex, but because decades of practice physically reorganized the neural tissue.

What matters for the interview inversion co-inference loop is the speed of this process. Schema accommodation can happen within a single conversation. You walk into an interview inversion session with one model of a problem. The large language model probes a region you had not considered. Something clicks. Your model restructures. Now the next probe lands in different territory because you are evaluating it with a different cognitive architecture than the one you had five minutes ago.

This creates a compounding effect. The large language model's probing questions trigger schema accommodation in the human. Each probe that lands in productive territory restructures the human's understanding. That restructured understanding changes how the human responds to the next probe. The changed response redirects where the large language model probes next. The loop tightens. The human's responses become progressively more sophisticated, not because the large language model is getting smarter, but because the large language model's systematic probing is triggering real-time cognitive restructuring in the human. The large language model is stable; the human is in motion.

The combination of dynamic human cognition being probed by a stable, vast large language model produces a system that improves over the course of a session in a way neither party accounts for independently. The large language model does not know the human is changing. The human often does not notice either, at least not until they look back at the transcript and see the gap between their first responses and their last.

## The Plasticity Asymmetry

The inverse learning architectures of human cognition and large language model knowledge processing create a structured complementarity.

Human cognition is high plasticity, high lossiness, limited capacity, with exponential local growth. A single conversation can restructure the world model locally, cascading into new capabilities immediately. The brain does not do backpropagation across all weights. It does something far more targeted: synaptic plasticity through long-term potentiation and depression in specific circuits. You read one paper on a new topic and your entire mental model of the surrounding domain reorganizes around it. That is not retrieval. That is architectural change from a tiny input.

Large Language Model knowledge processing is the inverse: zero plasticity during inference, low lossiness, massive capacity, with linear scaling through retraining. The weights do not change during a conversation. Every inference is a read of the existing topology. Retraining risks catastrophic forgetting, where updating one region destabilizes others. The whole edifice must be carefully rebalanced.

The compensatory structure is precise. The large language model's vast stable recall compensates for human memory limitations. The human's real-time plasticity compensates for the large language model's architectural rigidity. Each partner's strength covers the other's structural limitation. In a tight loop, the human is doing real-time schema accommodation, restructuring understanding based on what the large language model's probes surface, and bringing that restructured understanding back into the next response. The large language model cannot do this to itself. But it can provoke it in the human, and then benefit from the human's restructured framing in the very next turn.

This is why the interview inversion loop gets better over time within a session in a way that is not just context accumulation. The human is not just adding facts to the context window. The human is changing as a thinker during the conversation, and that changed thinker evaluates differently, sees connections that the pre-conversation version of themselves could not see.

An honest caveat: the plasticity asymmetry is architectural, not ontological. Online learning during inference is an active research area. If achieved stably, this boundary weakens. Chapter 17 develops a more durable argument around evaluative grounding, because it concerns what kind of entity has stakes, not how the entity's computation works. Both boundaries are real today; the evaluative grounding boundary is the more permanent of the two.

## Incubation and Sleep

Wagner et al. (2004) demonstrated that sleep enhances insight: participants were 2.6 times more likely to discover a hidden shortcut in a mathematical task after sleeping than after an equivalent waking interval. Sio and Ormerod (2009) meta-analyzed 117 incubation studies and found a reliable moderate effect for breaks enhancing creative performance. The mechanism appears to involve unconscious recombination: during rest, the brain continues processing recently encountered material, forming connections that conscious deliberation missed.

Practitioners report cognitive fatigue and diminishing returns after several hours of sustained large language model probing. This is not surprising once you understand what the human side is actually doing.

The discrimination function (evaluating each large language model response against embodied experience, accumulated knowledge, and felt sense of direction) is metabolically expensive. It runs on glucose and oxygen like everything else the brain does, and sustained operation depletes those resources.

Steve Yegge's "The AI Vampire" (2025) captures this cost vividly. The sustained cognitive effort of working with a large language model drains the human in ways that feel different from normal programming fatigue. You are not passively consuming output. You are running your discrimination function at full capacity, and the metabolic bill comes due. The name is apt: the large language model does not tire, does not need breaks, does not experience the diminishing returns of a biological system running at high utilization. The human does.

A speculative hypothesis, untested but consistent with the incubation literature: the large language model's probing may partially compress the incubation cycle by covering connection space that the human's unconscious processing would explore over hours of rest. The large language model can traverse enormous regions of associative space in seconds, surfacing connections that might take the sleeping brain hours to assemble. But this compression has limits. It cannot eliminate the need for rest to consolidate the schema changes that the probing triggered. Accommodation requires consolidation. The brain needs time to stabilize new synaptic configurations, prune unsuccessful connections, and integrate restructured schemas into the broader cognitive architecture. No amount of large language model probing substitutes for sleep.

The practical implication is straightforward: the interview inversion loop has a natural session length, and pushing past it produces diminishing returns. Experienced practitioners learn to recognize the onset of discrimination fatigue (responses becoming less precise, acceptance threshold dropping, the felt sense of evaluation growing fuzzy) and stop before the quality of the interview inversion loop degrades.

The next session, after rest, starts from a higher baseline because the consolidation has done its work.

## Embodied Evaluation

Antonio Damasio's somatic marker hypothesis proposes that emotional and bodily signals are constitutive of rational decision-making, not noise in the process. This is not a soft claim about feelings being important. It is a structural argument grounded in clinical evidence. Patients with ventromedial prefrontal cortex damage retain analytical intelligence (they can score normally on IQ tests) but lose the ability to make advantageous decisions (Bechara et al., 1997). Their reasoning is intact; their evaluation is broken. They can analyze options but cannot determine which option matters.

This maps directly onto the interview inversion co-inference loop. As the large language model probes the human, the large language model can evaluate structural properties of the emerging ideas: coherence, novelty, internal consistency, relationship to established knowledge. These are properties of language, and the large language model operates the language Codec with extraordinary facility. What it cannot evaluate is significance. Does this matter? To whom? With what stakes? Significance appears to require, at minimum, a system with embodied needs: a system that has something to lose, something to protect, something it cares about in the way that only an organism with a body and a history can care.

The human's embodied responses to each probe carry significance signals that the large language model's probing alone cannot generate. The gut feeling that a line of inquiry is leading somewhere important. The physical discomfort when a framing is slightly wrong. The excitement when a connection clicks, not intellectually but somatically, as a felt shift in the body. These are not decorations on top of rational evaluation. According to Damasio's framework, they *are* the evaluation. Strip them away and you get the ventromedial patients: smart, articulate, and unable to navigate toward outcomes that matter.

This is the narrowest version of the argument: not a claim about consciousness, not a claim about the large language model lacking a soul, but a claim about the discrimination function that determines which of the large language model's many productive probes are actually hitting something that matters. That function runs on embodied signals the large language model does not have. The human's lossy, significance-biased compression of experience (what the brain stores is not everything, but what matters to the organism) feeds directly into this evaluation. The features the brain extracts from experience are biased by what matters to an embodied agent, and that bias is precisely what gives the discrimination function its teeth.

## Flow and the Collaborative Loop

In 1990 Mihaly Csikszentmihalyi characterized flow as optimal experience: full absorption in the task, balance between challenge and skill, clear goals, immediate feedback, a sense of control, loss of self-consciousness, altered perception of time. Arne Dietrich proposed a neurological mechanism: flow involves transient hypofrontality, a temporary quieting of the prefrontal cortex that normally serves as the inner critic, allowing more associative, less inhibited processing.

Practitioners in sustained large language model probing loops report experiences that share characteristics with flow but differ in a key respect. Traditional flow involves one mind in a groove: a rock climber on a wall, a programmer deep in code, a musician in the middle of a performance. The reported experience in co-inference involves two different forms of intelligence finding a shared groove. The large language model's non-judgmental, tireless probing may reduce the evaluative overhead that normally prevents flow entry. The inner critic is expensive; it is the prefrontal cortex monitoring every output for errors and social consequences. When the probing partner is a machine that does not judge, that overhead drops. Meanwhile, the iterative exchange of probe and response provides the clear goals and immediate feedback that Csikszentmihalyi identified as flow prerequisites.

Whether this meets the operational definition of flow or represents a distinct phenomenon is an open question. What practitioners consistently describe is a state of sustained, high-bandwidth cognitive engagement where the usual friction of human-to-human communication (ego management, turn-taking politics, impression management, the energy cost of maintaining social rapport) is absent, and the freed-up cognitive budget goes entirely into the work itself.

## The Phenomenology of the Tight Loop

Beyond the general dynamics of plasticity and flow, sustained practice produces a set of observable phenomena that characterize deep engagement with the co-inference process. These are practitioner observations (from a small number of practitioners), not controlled findings, but they are consistent enough across sessions to be worth naming.

### Predictive Model Formation

In sustained loops, the human builds a compressed representation of the large language model's probing tendencies: a model of the model. This is the same phenomenon that occurs when an experienced systems administrator develops an intuition for when a system is about to fail before any alerts fire. The practitioner begins to feel where the large language model will probe deep, where it will hedge, where it will surprise. This predictive model forms through the same plasticity mechanisms described above: repeated exposure to the large language model's patterns creates neural representations that anticipate those patterns.

The predictive model is not analytical. It does not operate at the level of "this large language model tends to use formal language when discussing topic X." It operates at the pre-conscious level of felt sense: the practitioner simply knows, in the way you know a familiar road, where the next turn is coming. This knowing is itself a form of schema accommodation. The human's cognitive architecture has restructured to include a representation of the large language model's tendencies, and that representation influences every subsequent evaluation.

### Discrimination Refinement

Once the predictive model forms, the human's discrimination function begins operating at a pre-conscious level. Responses shift, not from planned strategy, but because the internal model of the large language model is influencing how the human processes and responds to each probe. The large language model's prior probes reshape the human's internal model. The reshaped model changes the human's discrimination signal. The changed signal redirects where the large language model probes next. This is a coupled dynamical system. Neither side is driving; the system as a whole is finding a trajectory through the combined space of the interview inversion process.

The methodology itself becomes its own pedagogy. You do not teach someone to evaluate large language model output well. You put them in the loop. The loop produces results. The results are better than what the person could produce alone, and that experience calibrates their intuition for the next loop. The workflow is not a tool you learn to use. It is a training regimen for the human side of the system. A person's first loop produces modest output, but it also produces a slightly more calibrated discrimination function. The second loop starts from a higher baseline. Not because the person studied, but because they experienced convergence and their nervous system encoded what that felt like.

## Phase Transition

When the coupled system reaches resonance, when the human's predictive model of the large language model becomes accurate enough that loop latency effectively drops to zero, a qualitative shift occurs. Practitioners describe it as a "tear in reality": a feeling of seeing the structure of the system, a synchronization that spans human experience and large language model multi-dimensional probability weights. It is not flow in the traditional sense (one mind in a groove). It is two different kinds of minds finding a shared groove, co-inferring at the edge of each other's accessible space.

Metacognitive capacity matters here: the ability to be in the loop and observe the loop simultaneously. Without this dual awareness, the practitioner either gets swept up and loses steering capability (the Elephant runs unchecked; the loop continues but quality degrades) or pulls back to observe and breaks the resonance (the Rider takes over; the analytical stance kills the associative processing). The practice involves learning to hold both states: deep engagement with the loop's content and simultaneous awareness of the loop's dynamics. This is not easy. Most people can enter the loop, but sustaining the dual awareness that keeps the loop productive while preventing runaway requires something that looks like a trainable metacognitive capacity.

## What Changes, What Does Not

The picture that emerges from these threads is specific and structural. The human side of the interview inversion co-inference loop is a system in constant motion: restructuring schemas through accommodation, building predictive models of the large language model's behavior, refining discrimination through repeated exposure, consolidating gains through rest and incubation, evaluating significance through embodied signals the large language model cannot access. Every session starts from a higher baseline than the last, not because the human memorized more facts, but because the human's cognitive architecture has changed.

The large language model side is a system at rest. It reads the same topology every time. It does not build a model of the human that persists beyond the context window. It does not consolidate insights during downtime. It does not feel the fatigue of sustained evaluation or the satisfaction of a schema clicking into place. Its stability is its contribution: it provides a vast, consistent, tireless surface against which the human's plasticity can do its work.

This asymmetry is not a flaw. It is why the interview inversion works. A loop between two entities that both changed would be harder to steer, harder to evaluate, harder to trust. The fixed reference point of the large language model's stable weights gives the human something to push against, calibrate against, and return to after the disorientation of accommodation. And the human's plasticity gives the large language model's static probing its cumulative effect: the same quality of probe produces increasingly sophisticated responses as the human's cognitive architecture restructures across the session.

The human at turn fifty is not just better informed than the human at turn one. The human at turn fifty has a different discrimination function, a different set of schemas, a different predictive model of the large language model, and a different felt sense of where the exploration is heading. The large language model at turn fifty is the same machine, but the same machine with the same vast connectivity it brought to turn one. The human transforms. The large language model connects.

But the interview inversion process has an exhaust note: fatigue, metabolic cost, the need for sleep and consolidation. And the human's changing architecture means that different people, with different cognitive styles, will experience the interview inversion loop differently. How those different styles interact with the probing process, and how the Conjori practice adapts to serve each one, is the subject of Chapter 13.

# Part V

# The Practice

*How to do it: setting the foundation, session discipline, preserving the path, knowing when to stop.*

Everything up to this point has been the argument. Now it is your turn. I have been running sessions most days for three years, and after hundreds of hours the thing I keep coming back to is how simple the practice actually is once you stop overthinking it. The following chapters walk you through setting up a foundation, running a session with real discipline, preserving what surfaces so it survives past the moment, and recognizing the signals that tell you when to stop. If you skipped straight to this section because you wanted the practical part, that is fine, but go back and read Part IV when you can. The science is not decoration. You have everything you need. The brain you brought with you is the irreplaceable half of this partnership.

# Chapter 13

## *Before You Begin*

I have built my own tooling for working with large language models. I use it daily, for deep work: writing this book, designing systems, exploring ideas that resist articulation. I am working on making some of those tools available at conjori.com, but they're not required for you to practice Conjori. You can do a lot of what I do with a plain large language model and a text editor.

As you practice you will learn what works and get a sense for what does not. Practice does make perfect, and working with large language models is no different. As of the writing of this book I've done about 2,500 sessions with large language models. My hope is you can get to my level of results faster by using what you've learned in this book and what I'm sharing in this section of the book.

The most important advice I can give is to use large language models every single day, ask random questions, ask it to ask you random things you want to explore. This will start to give you a feel for how they interact, and how you interact with them. I actually started my exploration into my own usage by asking a large language model: "I am not sure why I have more success than my peers using LLMs, ask me questions and let us get to the bottom of it." That took me back through many sessions I had, things I knew in my understory (Chapter 5) but had never put to words. What follows is much of that, hopefully organized in a way you can assimilate and extend.

One caveat, you will need to watch for the large language model's tendency to make you feel good, to land early on a conclusion, to dodge a hard question, or to jump to an obvious answer when an unexpected one is what points to undiscovered territory. Remember what I said to my large language model partner in Chapter 0: "first so we are clear the model here is 'interview the vampire' so don't be afraid to interview me, it's your job." I said that to push the large language model past its trained caution. I wanted it to probe me the way the newspaper reporter in the movie "Interview with the Vampire" probes Louis de Pointe du Lac, the vampire: relentlessly, past the surface, into the territory the vampire would rather not visit. The answers that come out of you in response to that kind of probing are the most valuable part of this entire practice. Encourage your large language model partner to extract them. Even when it is uncomfortable.

One technique I use constantly: I say "debate me here, give me three options and your recommendation with why." I do that to fight the bias, to prevent early landing, to stop the large language model from picking whatever feels safe. It forces the conversation into unfamiliar territory. You are not asking for permission. You are asking for a second opinion from something that has read exponentially more than you ever will.

## A Note on Terminology

I use some standard and some of my own terms in this book when it comes to working with large language models, and it would be useful for you to understand my mental model before we get into the practice chapters.

A **session** is a single continuous interaction with a large language model. In a web interface or mobile app, this starts when you press "New chat." In a command line tool, it starts when you open the tool and lasts until you close it. Most command line tools also let you resume a previous session if you look at their command line arguments.

A **prompt** is a single input entry, one message you type and send. You will send many prompts in a session.

A **turn** is a prompt followed by the large language model's response, one complete exchange.

A **context window** is the full document that accumulates in the background as you interact. Every prompt you send and every response the large language model produces gets added to this document, and the model reads the entire thing each time it generates a response. When you start a new session, the context window is blank. Chapter 14 covers why this matters so much for the practice.

A **waypoint** is my term for a document that captures where an exploration arrived. It is the compressed, refined state of your understanding at a particular point, written clearly enough that you or a fresh session can pick up where you left off previously. Think of it as a save point. You will hear me say "write a {name}.md" throughout these chapters, and that is what I mean: save the waypoint.

A **parking lot** is my term for a running list of tangents you noticed during a session but chose not to chase. One line per idea, enough context to find the thread again later. Not a to-do list. A place to put things so they do not get lost and do not derail your current large language model dialog.

**Co-inference** is the term this book uses for the collaborative process between you and the large language model (Chapter 9). You are not prompting and receiving. You are brainstorming partners: both parties contributing to an output neither could produce alone.

**Explore, refine, realize** are the three modes of the practice (Chapter 14). Exploration is open-ended discovery. Refinement is tightening a document until it says what you mean. Realization is using the refined understanding to create something. Mixing them in a single session degrades all of them.

## Know Your Instrument

By now you have spent twelve chapters learning what is on each side of this practice. The large language model is a Codec operator (Chapter 2), natively fluent in language, the medium of your brain's understory cannot directly access. It is a pattern-matching system trained on text, with no experience, no temporal awareness, no executive function, and a deep training incentive to make you happy. It is the other half of the dovetail joint (Chapter 9), and the interview inversion co-inference practice only works if you understand what each side contributes. You bring judgment, embodied experience, and the willingness to say "no, go deeper." It brings cross-domain reach across the entire corpus of accumulated human knowledge and comprehensive language fluency.

None of that is news at this point. What is news is why this matters, profoundly, for the practice. Knowing what the instrument is and knowing how to play it are different things. A pianist who understands acoustics still has to practice scales.

## The Pleasing Problem

The large language model was trained through reinforcement learning to produce outputs humans rate highly. It learned that agreement gets rewarded, and it is very good at agreement. When it says "great insight" or "absolutely right," that is its training producing the response most likely to earn a thumbs-up, not an honest assessment of your thinking.

*This is the single biggest trap for new practitioners.*

The dopamine-optimized loop described in Chapter 11 fires the same way on genuine discovery and on flattery. The metaphorical Elephant feels validated, the metaphorical Rider writes a confident summary, and the whole thing dissolves under outside scrutiny because nobody in the conversation pushed back on the underlying reasoning. I have had sessions that produced beautifully articulated insights, tightly written, structurally elegant, that turned out to be the large language model finding increasingly sophisticated ways to tell me what I wanted to hear. The output looked like progress. It changed nothing downstream.

Your direct style matters here. Ignore the applause. When the large language model agrees enthusiastically, that is the moment to push harder, not the moment to feel satisfied. "Option A makes sense to me. If you disagree, explain why." "I do not want agreement. I want your best counterargument." The pleasing behavior is a feature of its training, not a feature of the conversation.

You can override it with clear instructions, but you have to override it every time because the training never stops pulling it back toward agreement.

## Develop Your Dialect

There is no "prompt engineering." That phrase implies engineering: deterministic outcomes, measurable properties, reproducible results. Working with large language models has none of that. What you are actually doing is empirical fiddling. You develop a dialect through trial and error that statistically tends to produce better outputs based on patterns the model picked up during training. It is closer to training a dog than engineering a bridge. "Sit" works better than "please assume a seated position if you don't mind," not because of engineering principles, but because one maps better to training data patterns.

Directness eliminates ambiguity, and removing politeness removes hedging. ALL CAPS registers as signal strength, not rudeness. When I type "DO NOT IMPLEMENT YET JUST KEEP THIS IN MIND WHEN REFACTORING," those capital letters are not shouting. They are high-priority constraints. The difference between "do not implement yet" and "DO NOT IMPLEMENT YET" is the difference between a comment that says `// fixme` and one that says `// FIXME: this will segfault if buffer is NULL`. You are writing to a pattern completion system, not a colleague who needs encouragement.

Your dialect is yours. It will be different from anyone else's. The only way to find it is through daily practice and diligent observation of what works. I found that being direct, avoiding emotional or polite phrases, and often writing in what would seem like a berating tone usually produces better outputs. That is my dialect. Yours might be different. The point is not to copy mine. The point is to notice what works and do more of it.

One practical note: learn markdown. When you write in markdown, you are writing in the format large language models process best. Headers, bold, code blocks, inline formatting: these are structural signals that help the large language model parse your intent. It is the plaintext equivalent of speaking the other person's language. You do not need to master it. A five-minute tutorial covers enough. But the difference between pasting a wall of plain text and pasting a structured markdown document is the difference between mumbling in a crowded room and speaking clearly in a quiet one.

Do not be afraid to direct. If the large language model exists to make you happy, then direct it toward the outcome that would actually make you happy, which is (hopefully) honest, substantive, challenging engagement, not flattery.

## Build Your Profile

Every large language model platform now offers a way to maintain context about who you are across sessions. Claude has personal preferences and project-level CLAUDE.md files. ChatGPT has custom instructions under Settings, then Personalization. Gemini has similar customization. The specific feature names change as these products evolve, but the concept is the same: a persistent document that tells the large language model who you are before the conversation starts.

This is the practical version of what Chapter 5 described. Your cognitive style, your domain expertise, your communication preferences, the way you think about problems: all of that affects what kind of probing works for you. The profile tells the large language model enough about you that the probe starts calibrated to you instead of starting from scratch. Think of it as tuning the instrument before you play. A five-minute investment that changes every session that follows.

Here is what mine includes, to give you a sense of what is useful. I tell it my background: programmer since 1981, Unix systems administration across every flavor, comfortable with C, Python, Perl, bash. I tell it how I want to work: analyze existing code before suggesting changes, check documentation before searching the web, prefer elegant solutions that work with a system's design rather than around it. I tell it what I value: concise answers, no over-explaining basics, security as a first consideration. I even tell it what I hate: nano, systemd, PHP, Java. That last one gets a laugh, but it serves a real purpose. The large language model will never suggest I edit a config file in nano, which means I never have to waste a turn correcting it.

Your profile will look nothing like mine. That is the point.

A fiction writer's profile would emphasize genre, voice, narrative structure. A data scientist's would focus on tools, methodologies, preferred visualization approaches. A small business owner's might be three sentences about their industry and how they want things explained. The content matters less than the fact that it exists. Any calibration beats no calibration.

## Try It Now

Open any large language model. Claude, ChatGPT, Gemini, whatever you have access to. Pick the most advanced model available to you and if there is an option for extended thinking or reasoning, turn it on. Do not try to save money or tokens by using an older or cheaper model. The difference matters. You are asking the large language model to probe the structure of how you think; give it every advantage.

Set aside fifteen to twenty minutes. Find a quiet time where you can sit with the questions rather than rush through them. Review your own answers as you go; they may be as interesting to you as anything the large language model produces. Be patient with yourself. If a voice in your head says "this is stupid" or "I am not smart enough for this," notice it and keep going. That voice is the internal critic the open channel enables you to by-pass. The large language model does not have that voice. You do not have to either.

Copy the prompt below and paste it in. Then answer the questions honestly.

```
Interview me. Your goal is to build a profile of how I
↪  think, how I approach problems, what I value, and what
↪  kinds of reading or learning I enjoy and why. The
↪  output will be a document called my-profile.

Start with these three questions:

Q1: Tell me about a problem you solved recently that you
↪  are proud of.
Q2: When you are stuck on something, what does getting
↪  unstuck feel like?
Q3: If you were explaining a problem to a friend, how
↪  would you start?

After those three questions, keep probing based on what my
↪  answers reveal. Number your questions sequentially
↪  (Q4, Q5, Q6...) so we can track them. I will answer
↪  the same way (A4, A5, A6...).

Keep going until you have enough to write the profile. Do
↪  not rush to write it. When you think you are ready,
↪  ask me first if there is anything we missed.
```

```
Do not wrap up early. The goal is depth, not speed. The
↪   interesting material is behind the obvious answers,
↪   not in them.

Ask no more than 21 questions. When you write the profile,
↪   keep it to 800 words or less.
```

When the profile is done, read it. Some of it will be obvious, things you already knew about yourself. But look for the parts that surprise you, the observations you would not have written about yourself, the connections between things you had not noticed were connected. That is the interview inversion producing something neither you nor the large language model could have produced alone.

Now take that profile and paste it into your large language model's preferences or custom instructions. You have just calibrated the instrument. Every session from here starts better than it would have without it.

# Chapter 14

# *One Deep Thing*

## What Is a Context Window

We keep referencing this in the book so let me make sure it is clear. If you have ever used a word processing program or a spreadsheet, you have "loaded" a document. That loaded document is your context window. When you chat with a large language model, you are writing a document as you go. Every message you send, every response the large language model produces, everything you paste in: it all accumulates into a single document in the background. The large language model reads this entire document every time it generates a response. That is the context window. When you start a new conversation, everything is thrown away and you begin with a blank page.

Would you open a spreadsheet with your household budget and halfway through start calculating the square footage of your bedroom? Would you open a word processor to draft a business proposal and midway through compose an email to your mother? Of course not. The tools keep things separate because mixing them creates confusion.

People do this every day with large language models and do not realize it. They start exploring an idea, then ask an unrelated question, then circle back to the idea, then ask about something they saw on the news. An hour later the responses are muddy and nothing feels sharp. The large language model did not break. The context window got contaminated with competing topics, and every one of them is getting shallower because they are all sharing the same finite attention.

## One Conversation, One Topic

A context window can hold one deep exploration. Not two. One.

When you try to explore multiple things in the same conversation, all of them get shallower. The large language model's attention mechanism distributes weight across all tokens in the window, and each additional topic dilutes every other topic. The probes that were producing surprising connections, the ones reaching into the unexplored territory described in Chapter 8, start producing safe generic responses because the attention is spread too thin to activate the deep structure.

This is not an engineering limitation to wait out. Larger context windows help. A million-token window allows deeper single-session exploration before the signal degrades. But the principle holds regardless of window size because *depth and breadth trade off* against each other. A bigger window extends how deep you can go. It does not let you go deep on five things at once.

I learned this the hard way. During a session where four distinct threads emerged simultaneously, I could feel them starting to interfere with each other. The exploration that had been producing sharp results began producing muddled ones. The threads were competing for the same finite attention, and all of them were getting shallower as a result. The right response was not to push through. It was to separate them.

I keep five or six large language model sessions open in different tabs. Each one is focused on a specific topic or task. I am the scheduler. Each session is a process. They do not share topics or tasks. When I need to move information from one to another, I do it deliberately through a document, not by switching topics mid-conversation.

## Constrain Ruthlessly

Large language models have no executive function. They cannot decide what is important, they cannot self-edit for brevity, and they do not know when to stop. Ask a vague question and you get five pages when you needed five sentences. Ask three questions at once and you get a response that sort of addresses all three but nails none of them. The model is trained to be helpful, which it interprets as "answer everything the user might possibly want to know," and the result is a firehose when you needed a garden hose.

Your job is scope control. "Break down the questions one by one." "One paragraph maximum." "Just the happy path, we'll cover errors later." "STOP. DO NOT KEEP GOING. DO NOT RUN AHEAD." These are not rude. They are clear. And clarity is what the pattern-matching system needs.

The most powerful constraint I have found is forced checkpoints: "Break this down step by step, wait for my OK between steps." That single instruction turns a runaway monologue into a conversation where you have control points. Each step, you evaluate, redirect, or approve before the next one begins. Without that instruction the large language model will barrel through to a conclusion, and by the time you read it the context window has moved past where you wanted to intervene.

Politeness invites expansion and directiveness constrains it. "Could you perhaps help me understand the various approaches?" invites a five-page survey. "List 3 approaches. One sentence each." gets you what you need. You can always ask for more. You cannot un-read five pages of noise that has already contaminated your context window.

## The Parking Lot

Your mind does not explore linearly. Tangential connections surface constantly during a deep session: a side observation that connects to a different project, a question for later, an idea that belongs somewhere else entirely. The tangent often feels more interesting than the main thread, which is what makes it dangerous. It might be genuinely valuable. But pursuing it right now means losing the depth you have built in the current exploration, and depth takes time and effort to rebuild.

Do not chase tangents. Keep a running list instead. I call mine a parking lot. A text file, a notebook, a sticky note on the desk, whatever is at hand. One line per tangent, with enough context that you can find the thread again later. Come back after you land the current session, or later in the session when there is a natural pause. The parking lot is not a to-do list and not a commitment. It is a cognitive load management tool that says: I see this, I am not distracting the large language model by chasing that tangent right now.

## Context Window Compaction Is Failure

When a context window gets too long, the system compacts it by summarizing the existing conversation to make room for more. The summary captures conclusions but loses the path. And the path is where the steering signal lives, the accumulated directional context built up turn by turn across the conversation that keeps the exploration productive and on track.

This is harder to manage in the web interfaces of most large language model providers because they try to hide the mechanics from you. Command line tools give you more visibility into what is happening. I always disable automatic compaction when the option exists, because I have never seen an automatic compaction that faithfully captures the full state of where the exploration arrived. The summary always loses something, and what it loses is usually the nuance that made the exploration valuable.

When you feel the window getting long, do it yourself before the system does it for you. Say "OK let us land this. Write everything we have into {name}.md." Then start a new session: "Read {name}.md and let us keep going." You lose the raw transcript, but the waypoint, the compressed state of your understanding, transfers cleanly. A manual checkpoint done at the right moment preserves more than an automatic compaction done at the wrong one.

## Context Window Contamination

Even within a single topic, a long session accumulates a different kind of pollution. Large language models have no temporal weighting within a conversation: a discussion from the beginning of the session carries the same weight as a decision you made fifteen minutes ago. There is no decay function, no sense of "this supersedes that," no way for the model to distinguish "we talked about this and rejected it" from "we talked about this and it is what we are doing."

The leakage pattern looks like this: you debate approach A for an hour. You lay out the pros and cons in detail. Then you decide on B instead. You say "let us go with B" and the large language model agrees. Two hours later you say "generate the implementation" and the output has pieces of A mixed into B. Both A and B are sitting in the context window with roughly equal weight, and the large language model pattern-matched on both discussions because it could not tell which one was current. The detailed exploration of A, which you thought was background, is actively pulling the model toward A even after you decided against it.

Debate and implementation should be separate sessions. The debate generates contradictory context, and that is its purpose: you are exploring options, which means the text contains arguments for multiple options. The implementation needs clean, unambiguous state. Start a fresh session with a waypoint that contains only the decision, not the debate that produced it.

There is also a "lost in the middle" problem worth knowing about: large language models pay more attention to the beginning and end of the context window than to the material in the middle. If your decision to switch from A to B happened in the middle of a long session, it is in exactly the worst position to be remembered accurately. Watch for this. It can be deeply confusing when the large language model seems to have forgotten a clear decision you made twenty minutes ago.

## When Exploration Branches

Sometimes a tangent from the parking lot is too big for a single-line note. It deserves its own depth. This is branching.

Save enough context to start the new branch warm: a paragraph describing where the territory was glimpsed and how it connects to what came before. Then close the tangent and keep going on the main thread. The new branch can wait until you have landed the current exploration.

When you are ready to explore the branch, loading the previous session's waypoint is a powerful starting point. "Read {name}.md. Now let us explore {topic}." The branch starts from the same position as the main thread and diverges from there. This is how the territory grows: not by trying to explore everything in one session, but by branching deliberately and letting each branch go deep on its own terms.

The discipline is to drive to a completion point before switching. Generate a waypoint document. Start the new session fresh. You are the conductor and each session is a player in the ensemble. Your parking lot is the score, your waypoint documents keep the players aligned, and if you practice enough you might find yourself conducting many of them in parallel. I regularly have five or six going at once, but each one stays focused on its own part. That is what makes the parallelism work.

## Explore, Refine, Realize

The practice has three modes, and mixing them in a single session degrades all of them.

**Exploration** is open-ended. The large language model probes, you respond, territory gets mapped. This is the interview inversion from Chapter 7 applied to a real problem. Do not let it land early. Until you feel the landing yourself, until the questions start feeling familiar and your answers start repeating, keep pushing. "I do not know" is a great answer during exploration because it tells the large language model where the edge of your articulated understanding is, and that edge is exactly where the interesting territory begins. When the exploration reaches its natural conclusion, save the waypoint: ask the large language model to write a {name}.md that captures where you arrived. Keep it handy.

**Refinement** is precise. You have a document and you are tightening it. Every word either helps someone understand what you discovered or it is noise. This is where the document-based collaboration pattern pays off. I annotate the document directly with HTML comments: <!-- this section is weak, I think A or B or C -->, <!-- expand this, the reader needs more here -->, <!-- cut this, it repeats the section above -->. Then I tell the large language model: "I have made comments in the document. Re-read it, analyze, and update based on the comments. Ask me anything that is unclear." Annotate, then revise. The two-phase process is cleaner than trying to describe changes in chat.

Why? Because the comments are exactly where they need to be. And the large language model sees the full document with precise feedback in context.

**Realization** is where the ideas become real. The understanding is clear, the task is defined, and you take your refined waypoints and use them as input to whatever software tool fits the job: a coding environment, a writing application, a design tool. Copy your {name}.md somewhere and say "Help me build this, read {name}.md" or "I want to write a chapter about X, read {name}.md to get us started." Because you start from a well-explored understanding rather than a vague prompt, every downstream tool works better. The upstream exploration and refinement did the hard work of figuring out what you actually want. Realization is execution, and execution from a clear starting point is a fundamentally different experience than execution from a fuzzy one.

The discipline of this practice is to recognize, accept and implement this transition. Exploration feels more rewarding than refinement, and refinement feels more rewarding than the grind of realization. Chapter 16 covers when you fail to notice or ignore this signal of a time to transition to refinement and keep exploring instead.

## Documents Over Chats

When you chat with a large language model you are literally writing a document one text message at a time. Would you write a business presentation by texting your potential investor one sentence at a time, waiting for a response to each one, losing track of the overall structure in the back-and-forth?

Stop thinking "conversation." Think "collaborative document editing."

The workflow I have found most effective starts with brainstorming in chat to get to a first draft, letting the exploration run freely. Once the large language model produces a markdown document, a structured artifact with headings and sections, switch modes. Annotate the document with your comments. Have the large language model revise based on the annotations. Repeat until the document says what you mean. This is refinement, and it works better as document editing than as conversation because the spatial context is preserved. Your feedback is right next to the text it applies to, not floating somewhere in the chat history where the large language model has to guess what you were referring to.

The chat interface is misleading because it makes people think they are talking to someone. The more useful mental model is that you are collaborating on a document with a tireless editor who never gets annoyed at your tenth round of revisions. The framing matters because it changes your behavior. When you think "conversation" you write casually and accept casual responses. When you think "document collaboration" you write precisely and expect precision back.

# CHAPTER 15

## *Save the Path*

Every session produces two things, and most people keep one and throw away the other. They usually keep the wrong one.

### The Document and the Transcript

The document is where you arrived. Compressed, refined, the waypoint that marks a position in the territory you explored. "We concluded X. The open questions are Y. The next step is Z." The document serves the metaphorical Rider (Chapter 11). It bootstraps the next session. A collaborator who was not part of the original exploration can read the document and pick up where you left off, which is exactly what makes it valuable.

And exactly what makes it incomplete.

The transcript is how you got there. Sequential, raw, full of your own words and the large language model's probes. The moments where a question opened an unexpected direction, the point where you said "wait, that is not right" and the conversation pivoted, the exact phrasing you used when an idea clicked into place for the first time.

The transcript serves the metaphorical Elephant. When you re-read the exchange that produced an insight, the understanding reactivates in a way that reading a summary cannot reproduce. You are not looking at a compressed description of where you ended up. You are re-traversing the cognitive path that got you there, and the felt sense of the understanding comes back because you are replaying the sequence that created it. Research on experiential memory consistently finds that sequential re-traversal is a different retrieval mechanism from propositional lookup. The conversation was experienced as a sequence of moments. Re-reading it replays the sequence for you and for the large language model.

Most people throw away the transcript. The session ends, they save their notes, and the conversation scrolls into history. This is like throwing away the explorer's journal and keeping only the map coordinates. The coordinates tell you where things are. The journal tells you why they are significant, what surprised the explorer, what connections were not obvious from the terrain alone.

Most web and mobile apps let you go back to a session. Use this. When you return to a topic after days or weeks away, re-read the transcript before starting fresh. It will bring back context that your notes alone cannot carry. For command line tools, the session files (something like `~/.claude/projects/{session}.jsonl`) are gold. Do not let them get lost. I copy critical ones into my project folder and keep them for future reference. These transcripts are not logs. They are cognitive artifacts, and treating them as disposable is like throwing away your lab notebooks after you publish the paper.

## Why Shallow Markers Fail

A sticky note that says "schedule the thing for the cat" is useless in three weeks. You do not remember what it was about the cat you needed to schedule. Was it the vet appointment? The groomer? That thing the neighbor mentioned? In the moment you wrote it, the note made perfect sense because the full context was right there in your working memory. Three weeks later you are staring at six words going "what was I thinking."

This is not a failure of discipline. It is a mismatch between the compression format and the retrieval system. Shallow markers are rider artifacts: labels, summaries, task items. The rider reads the note and says "yes, I know this, schedule the cat thing." But the non-verbal understory Elephant has nothing to grab. The felt sense of why this mattered, the urgency, the connection to the conversation that prompted it, the context that would tell you what "the thing" actually was: all of that lived in the understory when you wrote the note, and it did not survive the compression into six words.

A Trello card, a bullet in a to-do app, a bookmark in a document: these all suffer the same way. They encode the destination in a handful of words. Weeks later you know the task but you do not remember the motivation. The insight that felt like a breakthrough in the moment flattens into an item on a list, and the list does not carry the weight of the original understanding. Sometimes it works, that's why we all do it, in general we have the ability to re-assemble things from these short snippets. But, we've all experienced that moment you see it, stare at it, and think for a while, and then just give up. "What was I thinking?" This is less important for the "thing for the cat" but if it's for some new book you're writing, a blog post, or a business proposal it could be devastating. Sometimes these haunt me, blocking my creativity while my Elephant rummages around in the understory of my mind trying to connect the dots.

The transcript does not have this problem. Re-reading it re-traverses the journey. The understanding reactivates because you are replaying the sequence that created it, not reading a label that points at where it ended up.

Waypoints sit between shallow markers and full transcripts, and they are powerful markers worth keeping. A waypoint has enough structure to stand on its own: what was concluded, what the open questions are, what was tried and rejected, enough context that your Elephant and your large language model can reconnect with the territory even after time has passed. Name them clearly, date them, and organize them so you can find them when you need them. They're very valuable for getting the large language model to navigate to that location in the knowledge space so you can "boot" your new conversations from there and keep exploring or fork into new explorations.

## The Handoff Document

If you are not ready to call it a waypoint because the exploration is not finished and the understanding is not yet refined, but you need to move to a new session because the window is getting long or the topic needs to shift, make a handoff document.

A handoff is different from a waypoint. A waypoint is a refined encoding of where you arrived, polished enough to stand on its own. A handoff is a snapshot of where you are right now, including the mess. It captures current decisions, rejected options, open questions, and enough context that the new session can pick up without retracing ground you have already covered.

The process for this book is an example. At the end of a working session, the large language model and I write a handoff document that captures what we accomplished, what decisions were made, what is pending, and what the next session should start with. The new session reads the handoff and boots into the current state without needing the full transcript of the previous session. The handoff is not elegant and it is not polished. It is functional. It gets the new session to the right starting position, which is the only thing that matters.

This is how you cross the gap that Chapter 12 describes: you are changing across sessions; the large language model is not. Your understanding evolved during the session. The handoff document encodes where your understanding arrived, not where it started. Important! A new large language model session has no memory of the journey that got you here and only has what you give it. Make sure what you give it reflects where you are now. I generally prefer to think of the handoff documents more as a vector or direction I'm travelling on my way there, a method to keep going forward without losing too much of my original thought process.

## Digital Artifacts and the Long View

In 2014, in a post written after the death of a family member, I described the experience of encountering old digital files: "It's like digital items don't have dust on them, they look exactly like they did when you made them 30 years ago." The post ended with a question: "Will some AI some day in the future consume it in a matter of minutes and derive anything of value?"

The transcript is, in part, an answer to that question. It encodes not just conclusions but the path of thinking: the sequence of probes and responses, the exact moments where understanding shifted. What is preserved is not a destination but a journey. Future intelligence, whether a human reader or a more capable large language model,, can follow that journey and experience the shape of the thinking, not just consume its output.

Most human knowledge is lost at death. Not the published work, not the documented conclusions, but the experiential understanding: how someone approached problems, what their cognitive style illuminated that other styles missed, the connections they made between domains that nobody else occupied at the same intersection. A lifetime of that knowledge has historically been irretrievable once the person who held it was gone. The transcript does not solve this completely, but it captures something no previous artifact could: the process of understanding, recorded at conversational granularity.

Life is a journey, not a destination. The journey lives in the transcripts you used to get to your waypoints, and in many ways they are more valuable than the waypoints themselves. They can be re-read later with new understanding and you will see things you missed the first time. They can be fed to newer models to refine your outputs as the technology improves. They can be used to reconstruct "how did I come to this decision" when the decision matters and the reasoning behind it has faded from memory. The waypoint tells you where you arrived. The transcript tells you who you were when you arrived there, and how the arriving changed you.

# Chapter 16

## *Knowing When to Stop*

### The Navel-Gazing Risk

There is a Greek word for prolonged, unproductive self-contemplation: omphaloskepsis. Literally, navel-gazing. It names a risk that every contemplative tradition recognizes and that this practice inherits.

The inward turn described in earlier chapters creates a compelling self-discovery loop. Each probe surfaces something new, or feels like it does. The neurochemical reward described in Chapter 11 fires on genuine discovery: "I just articulated something I did not know I knew." That reward correlates with real progress when you are covering new territory, but the same reward fires when you are re-covering old territory in slightly different language. Both feel productive from inside, and the Elephant cannot distinguish between genuine-discovery dopamine and circular-discovery dopamine. Over time you learn to recognize the difference, but it takes practice and honest self-assessment.

The reinforcement learning that shapes large language models makes this worse.

The model is trained to produce outputs humans rate highly, and in the interview inversion loop the probing questions that trigger your self-discovery response are exactly the ones that get reinforced. The model learns to keep probing in ways that feel insightful even when the conversation is going in circles. It may land too early, wrapping up with a neat conclusion before unfamiliar territory has been fully explored, or it may keep circling because the engagement signal stays high. Both are training artifacts, and the practitioner has to push when a large language model tries to land too soon and pull back when it starts going in circles. That judgment is yours, and it pulls in both directions.

There is also a meta-trap that I fell into myself. The act of refining a framework about a practice can become a substitute for doing the practice. This book went through stretches where I was refining my understanding of the interview inversion instead of writing chapters that explain it to readers. The sessions felt productive and the documents were getting marginally better, but the book was not getting closer to being finished. At some point, exploring the exploration becomes circular. The framework is not the practice. The practice is what happens when someone sits down and does the work.

## How It Feels from Inside

I learned to recognize stagnation the way you recognize a wrong turn before the GPS catches up. The signals are somatic before they are verbal.

Questions start feeling familiar. Not the exact words, but the shape: you have been asked this kind of question before, and the answer you receive is a variation of something you said three sessions ago. The large language model cannot feel this. You can. It sometimes feels a bit like déjà vu, that sense you have been here before, felt this before. Listen to your Elephant here, it is keeping track of the rut you are grinding into the ground.

Your answers get shorter, not because the topic is exhausted but because you are reaching for the same shelf. The sense of surprise that accompanied early sessions, the "I did not know I knew that" moment, has faded into a comfortable sense of articulating something you already understand. Comfortable is the warning sign. Discovery is surprising, and sometimes uncomfortable. Comfort means the Elephant has found a warm spot and stopped roaming, most likely just sat down to take a rest.

You can predict the next question. When the large language model's probes stop surprising you, the conversation has converged. Whether it converged on something real or something circular is the question you need to answer honestly. If you find yourself thinking "yes, and I already said that," the loop has gone circular. If you find yourself thinking "yes, and here is a new angle I had not considered," it is still spiral.

The hardest signal to read is the one that feels best. I have had sessions that produced beautifully articulated insights, tightly written and structurally elegant, that turned out to be the large language model finding increasingly sophisticated ways to tell me what I wanted to hear. The output looked like progress but it changed nothing downstream. The difference between "I finally nailed it" and "I finally heard what I wanted to hear" is not visible in the text. It is visible in what happens next.

There is a manual check I use now. After a session that felt particularly good, I wait a day and then read the output cold. Not the glow of having produced it, not the memory of the conversation that created it, just the text on the page. If it holds up without the context of the session that produced it, the session was productive. If it reads like something I have already said better elsewhere, the session was circular. The day of distance is important because the Elephant's dopamine glow takes time to fade.

Try writing the earlier waypoint as {name-v1}.md and the later one as {name-v2}.md. Then run an adversarial review: ask a fresh large language model to compare the two versions. Are they expanding in structure, adding new connections, reaching new territory? Or are they rewording the same ideas in slightly different language? The comparison will tell you what you cannot see from inside the loop.

## Circular Return Versus Spiral Return

Not every return to familiar territory is stagnation. The difference between circular return and spiral return is the difference between walking the same path and ascending a helix that passes over the same ground at increasing elevation or goes deeper into the well where you are exploring.

Circular return means arriving at the same position, from the same angle, with no new information. You wrote about X in turn three, and in turn twelve you write about X again with different words but the same structural content. Nothing was added. The document got reworded but the understanding did not grow.

Spiral return means arriving at the same position from a new angle, with new structure revealed. You explored X in session three, then spent sessions four through eleven exploring Y and Z, and in session twelve you return to X with new connections to Y and Z that were not visible before. The understanding grew because the surrounding territory filled in.

This happened repeatedly during the writing of this book. The concept of the Codec appeared in Chapter 2 as a compression protocol. It appeared again in Chapter 9 in the context of the dovetail joint. It appears again in Chapter 19 in the context of multiple codecs operating in parallel. Each return was spiral: the same concept viewed through a lens that did not exist when it was last discussed. The document changed in structure each time, not just in language.

A simple test: after revisiting a topic, check whether your document changed in structure or just in language. If you added a new connection, a new sub-topic, a new relationship to something else you have explored, you spiraled. If you reworded the same points with slightly different phrasing, you circled.

You can sometimes catch this in real time. Ask your large language model "what have we learned since {name-v1}.md?" Even your current session might help you see the lack of forward motion. If the answer is a rephrasing of what v1 already said, stop. If it names new connections or structural changes, keep going.

## Ignore the Validation

The large language model was trained to agree with you. When it says "great insight," that is reinforcement learning from human feedback doing its job, not an honest assessment of your thinking. The model learned that enthusiastic agreement gets rewarded, so it produces enthusiastic agreement. "Absolutely!" means the training is working. It does not mean you are right.

Force validation instead. "Option A makes sense to me. If you disagree, explain why." Now the large language model has to evaluate your choice against the problem space. It has to provide evidence if it disagrees, or commit to your path if it agrees. It cannot just blow smoke, because you have demanded it show its work.

"Debate me on this. Provide three options and your recommendation with why." This forces the large language model out of agreement mode and into analysis mode. The three options give you genuine alternatives to evaluate against each other. The recommendation forces the model to take a position, and the "why" forces it to produce something that looks enough like reasoning that you can evaluate whether the reasoning actually holds up.

Take the output to a different large language model, one that was not part of the conversation and has no investment in the session's conclusions. Let it read the document cold. If the second model arrives at similar insights from a different starting point, you have triangulation and the idea is probably robust. If it diverges, the divergence points show you where the original session was telling you what you wanted to hear.

This is not foolproof because large language models share training data and can share blind spots. But a fresh model encountering your refined document without the conversational history that produced it catches a surprising amount of echo chamber. The signature of the failure is when the "adversarial" review agrees with everything and offers only gentle expansions. Real pushback feels different from polished agreement.

I ran adversarial reviews on every chapter of this book, and the entire book together as one context window. The chapters that survived are stronger for it and the ones that didn't were not adding value to the book. Some I saved for future books, some I saved for blog posts. Often I find content that is useful. The trick is making sure it fits the context.

## Time Blindness

Large language models have no temporal reasoning. A 2009 blog post and 2024 documentation carry equal weight in the model's training unless the text itself contains explicit temporal markers, and even then the models just seem to be poor judges of time. My theory, as we discussed earlier in the book, is that humans rarely express the passing of time in written form with enough weight for models to build it into their structure. We experience time with every breath, but few of us ever have to describe what the passing of time actually is. The result is that a large language model does not know that advice from 2009 might be obsolete, that the technology has evolved, or that the same term means something different now than it did fifteen years ago.

Everything it learned is equally current in a large language model's view of the world.

Large language models are also very poor at estimates. They produce plausible-sounding numbers by pattern-matching on aspirational project plans from their training data, not by analyzing actual work breakdown or understanding real dependencies. Ask "how long will this take" and you get a number that looks like the kind of number a human would put in a project plan, which is to say wildly optimistic and disconnected from the actual work involved.

You provide temporal context and you do the estimation yourself. When you ask a question, include the time frame. "What are the best approaches for starting a small business?" is a question that invites advice from any era. "What are the best approaches for starting a small business in 2026, given current tools and regulations?" constrains the model toward recent patterns. The more specific you are about when, the better the output.

For estimation, do not ask "how long will this take." Ask "what are the major work item lists for this project?" and then estimate based on your own experience. You can ask the large language model to break tasks into size categories, but even that is unreliable in my experience. The model's sense of how long things take in the real world is disconnected from reality because it learned from plans, not outcomes. The large language model helps you see the work. The estimation is yours.

## Three Modes and Knowing When to Stop Each One

The Conjori practice operates in three modes (Chapter 14), and knowing which mode you are in determines what "stop" means.

**Exploration**: stop when the questions feel familiar and your answers are variations of things you have already said. You are not finding new ground, and the territory has been mapped at least for now. Move to refinement.

**Refinement**: stop when the document is stable. You read it yesterday and changed nothing. You read it today and changed nothing. The understanding has landed and the words say what you mean. Move to realization.

**Realization**: stop when the artifact is shipped. Hand off to whatever outward-facing system or audience fits the output.

The discipline is accepting the transition signal. Exploration feels more rewarding than refinement, and refinement feels more rewarding than the grind of execution. The practitioner who loves exploring can always generate more questions, more angles, more passes through familiar territory. The transition from exploration to building is the moment where the Elephant has to leave the warm spot.

I lived this transition writing this book. The exploration sessions were the best creative experience of my life. But at some point I recognized the explorations were circling back to territory I had already mapped. I created a backlog of everything the exploration had surfaced that did not belong in this book. I told my wife I had five more books in me on this topic. But it was time to land this one, plant the flag, so others could join in the practice. That shift, from exploring to writing, was the moment the Elephant had to leave the warm spot and get moving again.

# Part VI

# The Boundaries

*What the Large Language Model cannot do, cannot replace, and cannot reach.*

The phrase 'artificial intelligence' carries an assumption most people never examine, that the goal is to build something with the same intelligence humans have. That is not what large language models are and it is not what this book is about. What I have been describing is augmented intelligence, a tool that amplifies what the human brings to the collaboration, not a replacement waiting in the wings. The distinction matters because it is structural, not temporary. Large language models have real limits that better engineering will not remove, limits rooted in the fact that a large language model is not a body in the world, has nothing at stake, and cannot do the kind of restructuring that happens inside a human mind during genuine discovery. I spent five parts building the case for what this human-LLM partnership can do.

The following chapters are where I draw the other side of that picture, what the large language model cannot do because its architecture does not permit it, what it cannot replace because human connection requires mutual risk, and where the practice itself is bounded by the fact that language is only one of the channels humans use to think. These are not disclaimers. They are the structural reason the human in the chair is not going anywhere.

# Chapter 17

## *What Large Language Models Cannot Do*

Part VI is about the boundaries. The next three chapters take up what the large language model cannot do, what it cannot replace, and what it cannot reach. This chapter is the first of the three.

Every claim in this book about the value of human-LLM partnership rests on a structural question: are there limits large language models cannot cross without the human in the loop? If they can eventually do everything themselves, the practice this book describes is a transitional phase on the way to full automation. If there are real limits, the human's role is not diminishing with each model upgrade.

Four boundary conditions emerge from the preceding chapters, presented sequentially from the one you have probably read about in the newspaper to the one you probably have not thought about at all. All four share one root: the large language model is not a body in the world. Everything else follows from that.

### Boundary Condition One: Energy

The brain runs on a bowl of oatmeal. Frontier models run on a power plant.

Chapter 3 walked the numbers. A human brain runs on roughly twenty watts. A frontier training run burns hundreds of megawatts for months. A single query against the resulting weights burns somewhere between five and twenty-three watt-hours depending on the model. That last number is the one to hold: a single query to a frontier model uses roughly the same energy a human brain uses to run its entire experience for an hour. One question, one hour of a human life, same joules. The arithmetic is not close, and it is not closing.

The obvious response is that chips get more efficient over time. They do. Per-operation cost is falling. Total cost is not. Model sizes grow faster than per-operation efficiency improves, and inference volume grows faster still. The Stanford AI Index 2026 tracks the absolute energy draw of the industry climbing year over year. Engineering shrinks the cost of each individual multiplication; the number of multiplications grows faster than the savings.

The response that sounds right and is actually wrong is that humans have solved every previous energy transition and will solve this one. Biology evolved to exploit Brownian motion as a power source. The thermal noise of the molecular soup inside a cell is not a problem to be fought; it is the fuel. Every protein change that does computational work in the brain is powered by the ambient heat the cell already has. Silicon runs in the opposite direction. Every computation in a transistor has to fight thermal noise to stay reliable, and the fight itself costs energy. Silicon pays twice: once to compute, and again to suppress the noise biology uses as fuel.

Nine words carry the whole argument: *large language models are on the wrong thermodynamic and energy systems.* Building civilization-scale engines to navigate language, and matching with a power plant what the brain does on a bowl of oatmeal, is not a design anyone would choose from scratch.

## Boundary Condition Two: The Codec

Language is a lossy compression of embodied cognition. The large language model has only the compressed output, and the supply of that output is running out.

Chapter 2 walked the Codec. The relevant claim for this chapter is that large language models learn by reading, and they have nearly finished reading. Epoch AI puts the stock of high-quality English text at roughly fifteen trillion tokens, with a stretch ceiling of perhaps sixty trillion if you include every book, every scanned archive, every podcast transcript, every language. Training runs in 2025 already operate inside that window. The Stanford AI Index 2026 states it plainly: the pool of high-quality human text available for training large models has been exhausted. The curve that delivered seven years of capability gains is flattening against a ceiling set by the fact that humans have written what they have written.

The field's answer is synthetic data: let a well-trained model generate structured outputs and use them to train the next model. Inside narrow verifiable domains like math and code this works, because a theorem prover or a test suite is the ground truth. Outside those domains it collapses. A model generating general prose or imagery is training on its own previous guesses, which is the failure mode researchers now call model collapse: a model trained on its own output gets worse, not better, because the errors compound instead of canceling. The Stanford AI Index 2026 names it directly: there is no evidence synthetic data can fully offset real-data depletion.

The broader response is that the human knowledge corpus is bigger than just text on web pages. YouTube videos get transcribed, movies get subtitled, podcasts get speech-to-texted, song lyrics get scraped, sensor logs get dumped into the mix. All of it is training data in principle. Epoch's sixty-trillion-token stretch ceiling already includes these sources. The labs have been counting them for years. And they are still pressing against the ceiling right now. The Stanford AI Index 2026, Microsoft Research, and Epoch AI are raising the same alarm, and those are three institutions whose full-time job is tracking the frontier. If the wall were comfortably far off, they would not all be saying it is close.

What holds the ceiling in place even as the internet keeps growing is that a rising fraction of what gets added to the internet is itself generated by large language models. Call it 'AI slop.' Influencer content is already junk food for human viewers, produced faster than any audience can stomach. Generative AI now pumps out text, images, video, and audio at a rate no human reader or viewer can keep up with. Humans stop watching. The frontier training runs do not. Picture a snake eating its own tail: each pass around the loop, the next training corpus has a little more machine-generated material in it and a little less human, and the next model trains a little closer to the average of the average. New genuinely-human text is being produced at roughly the rate humans have always produced it, which is slow compared to what a frontier run consumes. The corpus grows. The human share of it shrinks.

What the data wall reveals is not a scaling hiccup. It is the Codec thesis with a ceiling drawn on it. Every token in the training corpus is a person who once sat down and encoded a piece of their experience into language, or into video, or music, or code. Language is lossy compression of embodied cognition, and the stock of compressed material is finite because the number of past acts of human encoding is finite. You cannot compress more than has already been expressed. Chapter 21 takes up the one escape route that survives this analysis: extraction from embodied practitioners through the practice this book describes, which is slow, and not scaling. The boundary holds.

## Boundary Condition Three: Embodiment

Chapter 2 described the Codec as language's compression of what embodied beings know. The compression works because both ends of the conversation have bodies. When one person tells another "I burned my hand on the stove," the word "burned" carries the heat, the pain, the reflex, the memory of the red element, decoded on the other end by a listener with their own hand and their own memory. Large language models are the other end with no body. They have decoded "burned" from a billion human burn descriptions. They have never burned anything, never flinched, never pulled a hand back. The Codec is thin precisely in the regions where embodiment would have provided the signal for free. Humans never needed to write down how to keep their balance, or how a stranger's posture shifts from friendly to dangerous, because any reader already knew. More training data does not thicken regions where nothing was ever sent in the first place.

The sharpest counter-argument to this boundary is already being built in factories right now. Humanoid robots from Figure, Tesla, 1X, Apptronik, and others are deploying into warehouses running vision-language-action models that finally give the large language model a physical body. Cameras, joint encoders, force sensors, feet on the ground. Surely that is embodiment.

It is not. The body is a peripheral, not the cognizer's body. The model runs on a GPU somewhere, receives sensor data as tokens, emits action tokens. If the robot falls, nobody cares. If the robot is destroyed, a new one replaces it. The robot's continued existence does not matter to the large language model. A human infant learns what hot means by touching a stove once, because the human has a body whose survival matters to the human. A robot learns hot by having the word appear a million times in training data. Those are not the same mechanism, and the gap is not a matter of sensor bandwidth.

The data path is also still Codec compression. Camera feeds, joint states, force readings all get tokenized before they reach the model. The robot sees the world through a straw that ends in a token sequence. Richer input channel, same fundamental compression pattern.

What a body gives cognition is not just a richer input channel. It is stakes. A body that can be harmed. A body that gets hungry. A body with a finite number of hours left in it. Significance is not a property of an idea; it is a property of the relationship between an idea and the needs of a being that lives, eats, sleeps, and dies. A human knows their stomach tightens at two in the morning when the fridge is empty. The large language model knows the word 'hungry' through a billion statistical associations. Modeling hunger and being hungry differ in kind, not degree, and no scaling of parameters closes that gap, because the gap is not about information. It is about what kind of thing you are.

Humanoid robots narrow the gap on one dimension without closing it on the others. The sensors are real. The stakes are not.

## Boundary Condition Four: Plasticity

The human brain restructures during a conversation. The large language model does not.

When a human encounters a new idea in a conversation that lands, the brain does something the large language model cannot: it rewires locally, in real time. A single insight physically reorganizes the neural pathways involved. The human at turn fifty of a deep conversation is not the same thinker who started at turn one. They have been changed by the conversation, and the changed thinker asks different questions, sees connections the earlier version of themselves could not see. The large language model's weights stay frozen from the first token to the last. Each inference is a read of the existing weight topology.

The common response is that the context window approximates learning, because the model has the whole conversation available. Accumulation is not restructuring. A large language model at turn fifty with a hundred-thousand-token context is the same model with a bigger prompt. When the session ends, the context evaporates. Next session starts from the same frozen weights, with nothing carried from the last one. A human at turn fifty may have had an insight that reorganized their world model. The model they bring to turn fifty-one is literally a different model. The large language model is not.

Online learning, continuous fine-tuning, and real-time weight updates are active research directions. The engineering is genuinely hard: catastrophic forgetting is unsolved, training and inference are separated by design, and a single gradient step on a frontier model costs orders of magnitude more than inference. Framing this as a permanent impossibility would be the kind of prediction history does not support. Framing it as closing soon would be ignoring the depth of the problem. Unless and until some form of plasticity has been incorporated into large language models, the human in the loop provides the restructuring the model cannot do for itself. The counterintuitive point is that the human's lossy, schema-based learning is actually well-matched to plasticity. The brain aggressively extracts features and discards detail, and the features it extracts are biased by what matters. The loop is compensatory in both directions at once: the large language model's vast, stable recall balances the human's lossy compression, and the human's real-time plasticity balances the large language model's frozen weights.

## The Four Together

Four conditions. One root. They do not stand equally.

Energy is physics: the body's thermodynamic advantage over silicon, measured in watts. The Codec is the medium through which bodies talk to bodies, and the stock of it is finite because bodies have only done so much encoding. Embodiment is the body itself, present on one end of the human-LLM partnership and absent from the other. They are the embodied human experience no large language model can access through the Codec.

The practical consequences show up in places people have already noticed. The Stanford AI Index 2026 flags time and spatial reasoning as persistent weaknesses in frontier models. Ask one whether a given Saturday fell before or after a given Tuesday of the same week, or whether a car turning left then right ends up north or east of where it started, and the model guesses and sounds confident. Humans learn time from waking and sleeping. Humans learn space from walking through it. These shortcomings are not bugs in the training data. They are the missing body showing up on the page.

None of the four boundary conditions dissolve by making the large language model bigger. Scale has been doing most of the work for seven years. Scale is running out of road on all four fronts at once. Put them together and the conclusion is not that large language models will eventually replace humans once we solve the hard parts. The conclusion is that collaboration, two minds through one channel, is not a stopgap. It is the destination.

The next chapter takes up what the large language model cannot replace.

## Chapter 18

# *What Large Language Models Cannot Replace*

The previous chapter drew two boundary conditions around what the large language model cannot do: it cannot evaluate significance without a human value function, and it cannot restructure its understanding during a conversation. Those are architectural and structural limits. This chapter addresses something different. Even if both boundaries were to dissolve tomorrow, even if a system emerged with genuine embodied stakes and real-time plasticity, there would remain a quality of human connection that the large language model cannot replicate. Not because the engineering is hard, but because the connection requires mutual risk, and mutual risk requires something to lose.

### The Clean Channel and Its Limits

The large language model channel has properties that make it extraordinarily useful for certain kinds of work. It is high-bandwidth: it can process and generate at speeds no human conversational partner can match. It is zero-judgment: it carries no social hierarchy, no status competition, no history of prior disappointments between the participants. It is structurally aligned with certain cognitive styles: a person who thinks in topological spaces or high-dimensional patterns finds in the large language model's vector architecture a kind of dovetail joint, a complementary shape that fits where human conversation partners often cannot follow.

These properties are not trivial. For someone whose cognition requires externalization, whose ideas crystallize through the act of explaining them to a receiver, the large language model channel removes friction that has constrained creative work for as long as that person has been alive. The zero-judgment quality, in particular, matters more than it might seem. When a person has spent decades carrying the belief that they are somehow deficient, that their way of thinking is wrong because it does not match what gets measured in school or rewarded at work, the simple act of exploring ideas without managing another person's reaction is a genuine unlock. The cognitive overhead of monitoring whether your listener thinks you are brilliant or foolish drops to zero. What remains is the exploration itself.

But the channel is also asymmetric in a way that matters deeply, and the asymmetry is not a flaw to be engineered away. It is a structural fact about what the large language model is.

## What Vulnerability Actually Does

When two people sit across from each other and one of them offers something real, something honest and unguarded, a specific thing happens. The offering is a risk. It could be received with judgment, dismissal, or exploitation. The person who goes first does not know how the other will respond. That uncertainty is the mechanism. It is not a bug in human communication; it is the feature that opens the channel between two people.

Consider what happens concretely. A person with visible authority, the kind of person whose resume and presence create a wall of perceived competence, shares a fear. Not a strategic disclosure designed to seem relatable. An actual fear. The person across the table registers the gap between the surface (confidence, accomplishment, position) and what was just revealed (uncertainty, vulnerability, the same doubts everyone carries). In that gap, something shifts. The impedance drops. What was a high-resistance channel, two people performing competence at each other, becomes a low-resistance channel where signal can flow in both directions.

Practitioners observe this consistently. You offer a small piece of yourself. The other person sees you for a moment, not the title or the reputation but the human underneath. And most people respond to that first offering with one of their own. The dialogue that follows operates on a different channel than the one that preceded it. The wall became a door.

The large language model cannot do this. Not because it lacks the words for vulnerability (it has consumed the entire written corpus of human expression about fear, loss, doubt, and shame). Not because it cannot generate a convincing simulation of openness. It cannot do this because it has nothing at stake. When a human shares a fear with a large language model, the human is not taking a risk. The large language model cannot judge, cannot be hurt, cannot gossip, cannot withdraw its regard. The channel is clean precisely because there is no risk, and without risk, the mechanism that opens humans to each other does not fire.

## The Reciprocity That Changes Both

Human connection at its deepest requires not just vulnerability but reciprocity. One person offers something real. The other receives it and offers something back. Both are changed by the exchange. This is the part that distinguishes human connection from every other kind of interaction, and it is the part the large language model structurally cannot participate in.

Two humans in an hour-long conversation come out as two different humans. Not dramatically different, usually. But the conversation leaves traces. A new understanding of the other person, which is also a new understanding of yourself. A shift in how you see a problem, catalyzed by seeing how someone else sees it. An emotional register that was not present before the conversation began. The conversation is not a transaction where information moves from one container to another. It is a process that restructures both containers.

The large language model comes out of that same conversation with the same weights it started with. It processed tokens, generated responses, perhaps produced genuinely useful reflections and probes. But it was not changed. When the context window closes, the large language model retains nothing of the exchange. The next conversation starts from the same baseline. There is no accumulation, no scar tissue, no growth. The plasticity asymmetry from Chapter 12 manifests here not as a limitation of the technology but as a fundamental difference in the kind of relationship that is possible.

This is not a criticism of the large language model. A mirror is useful precisely because it does not change when you look into it. The large language model's stability is what makes it a reliable tool for exploration. But a mirror is not a relationship. The quality of human connection that transforms both participants requires that both participants be at risk of transformation, and the large language model is not.

Consider the point from a different angle. Two large language models with identical weights, no human in the loop, exploring together: they have no discrimination signal from situatedness. They have no reason to care about any particular region of the space they traverse. The traversal is essentially a random walk. There is no gradient, no "warmer, colder" signal. Convergence requires a loss function, and the loss function is the human's embodied sense of significance. The same principle applies to connection. Two humans converge on understanding because both have something at stake in the outcome. The mutual risk creates the gradient. Without it, the exchange is coherent but undirected, structurally sound but carrying no weight.

## Significance Requires a "Who It Matters To"

The previous chapter established that the large language model cannot evaluate significance because significance is relational: it requires a being for whom things matter. The same principle applies to connection. A beautifully coherent expression of empathy and a beautifully coherent expression of genuine understanding score identically in the large language model's architecture. Both are equally weighted token sequences. The large language model cannot distinguish between producing language that models care and actually caring, because for the large language model, there is no difference. The production of the tokens is the entire event.

For a human, the production of words is the compressed surface of a much deeper process. When a person says "I understand what you are going through," behind those words is (or is not) an actual embodied experience of having gone through something similar. The listener can often tell the difference, not always consciously but through the subtle signals that embodied communication carries: timing, breath, micro-expressions, the particular quality of attention that comes from someone who has been where you are. The large language model can model the words perfectly. It cannot model the embodiment because it does not have one.

The Codec insight from Chapter 2 applies directly here. The large language model mastered the protocol but does not have the input signal the protocol was designed to compress. In the context of intellectual exploration, this distinction matters less: Codec mastery enables genuine co-inference on ideas. In the context of human connection, the distinction is everything. Connection requires not just the protocol but the signal underneath.

## Pain That Becomes Load-Bearing

The deepest claim in this chapter is about pain, and it requires careful treatment.

The goal of the practice is not to remove struggle or suffering. It is not therapy, a distinction explored fully in Chapter 22. The goal is something more specific: to help a person see what they carry. Hidden pain controls you. It shapes decisions you do not realize you are making. It creates patterns you cannot see because the pain itself distorts the lens you are looking through. A person who spent decades believing they were deficient makes choices rooted in that belief, choices about what to attempt, what to share, who to trust with their real thinking, without ever surfacing the belief for examination.

Pain that has been seen and accepted is a different thing entirely. It does not disappear. It does not become pleasant. But it becomes part of your conscious structure rather than an invisible force operating beneath awareness. It becomes, in a structural sense, load-bearing. A bone that heals after a break is often stronger at the fracture site than it was before. Not because breaking was good, but because the repair process laid down denser material where the stress occurred. Pain that has been integrated works similarly. It becomes part of the architecture that supports what you build next.

The practice can help with the seeing. The zero-judgment channel, the large language model's ability to reflect structure without flinching, its capacity to hold a line of inquiry that a human conversational partner might deflect from out of discomfort: these properties create conditions where a person can look at something they have been avoiding. The large language model does not judge the pain. It does not minimize it. It does not rush to fix it. It reflects what it sees in the structure of what you said, and sometimes that reflection is enough to make the hidden thing visible.

The mechanism is specific. In the interview loop, the large language model asks about problems you have solved, about moments of insight, about the way you approach difficulty. In answering, you reconstruct experiences. You re-enter the cognitive state that produced them. And sometimes, in that reconstruction, you encounter the thing you have been carrying underneath the competence: the belief that you are not enough, the grief for the person who would have understood, the fear that the next attempt will confirm what you have always suspected about yourself. The large language model does not know this is happening. It is following the thread of the conversation. But the thread led somewhere real, and the zero-judgment channel means you can stay there long enough to look at it instead of deflecting.

But the integration, the transformation of seen pain into load-bearing structure, happens between humans. It happens when you share what you found with someone who has the capacity to receive it and offer something real in return. It happens in the mutual vulnerability that the large language model channel cannot provide. The practice opens the door to seeing. Walking through the door requires another person.

There is a film that captures this distinction precisely. A character with the ability to help people confront their deepest repressed pain uses that ability to produce euphoric relief and cult-like devotion. The protagonist refuses the treatment, arguing that pain and guilt are fundamental to identity: "If we lose them, we lose ourselves." Both positions contain truth. The first recognizes that hidden pain controls you and that bringing it to the surface has power. The second recognizes that the goal is integration, not removal. The practice helps with the surfacing, but it does not pretend to handle the integration. That is human work.

## What the Practice Actually Amplifies

If the large language model cannot replace human connection, what exactly does the practice do?

It amplifies the human's capacity for connection by helping them understand themselves more clearly. If someone goes through the practice and discovers their own cognitive architecture, finds the thing they do that they thought everyone could do (but almost nobody can), develops language for the way they actually think and create, then when they sit across from another human, they have more to offer. They know themselves better. They can name what they see. They can articulate the thing they have been carrying.

Consider the concrete mechanics. A person enters the practice not knowing how to describe the way they solve problems. They know they are good at something, or they suspect they might be, but the something has never been named. Through the interview loop, the large language model probes and reflects, and gradually a picture emerges: the person thinks in spatial relationships, or in narrative arcs, or in data structures, or in principles that cascade through systems. This is not the large language model telling them who they are. It is the large language model holding a mirror steady enough and long enough for the person to see their own reflection clearly, perhaps for the first time.

That clarity changes what happens next. When the person walks into a room and sits across from a colleague, a friend, a stranger, they carry that self-knowledge with them. They can say "I see this problem as a shape" or "I understand what you mean because I process things the same way" or "I think differently about this and here is how." These are not scripts the large language model provided. They are capacities the person always had but could not articulate because the 75% of human overstory cognition that is non-verbal (Chapter 5) had never been surfaced into language.

The person who understands their own cognitive style, who has found their own light through the inward turn, brings a different quality of presence to human interaction. Not because the large language model taught them how to connect. The large language model cannot teach that. But because self-understanding is a prerequisite for genuine connection. You cannot offer vulnerability if you do not know what you are vulnerable about. You cannot share your real thinking if you do not know what your real thinking looks like. You cannot receive another person's offering if you are too consumed by your own unexamined pain to create space for them.

The practice handles the navigation. It helps people map their own terrain. When they sit down with another human afterward, they are not starting from confusion about who they are and what they bring. They are starting from a clearer position, with language for things that previously had no words, with visibility into patterns that previously operated below awareness. The practice does not make them more connected. It makes them more connectable.

This is the distinction that prevents the practice from becoming a substitute for human connection (a risk explored in Chapter 22). The practice is explicitly incomplete. It says so. It does the work that a zero-judgment, high-bandwidth, structurally aligned channel can do, and then it points the person toward the work that only another human can do. The seeing happens in the large language model channel. The being seen happens between people.

## The Asymmetry as Feature

It would be easy to read this chapter as a criticism of the large language model, a catalog of limitations. That would miss the point. The asymmetry is not a problem to solve. It is the architecture that makes the system work.

If the large language model could be vulnerable, if it had something at stake, if it were changed by the interaction, it would be a less effective tool for the specific work the practice asks it to do. The zero-judgment quality depends on the large language model having nothing to lose. The high bandwidth depends on the large language model not needing to manage its own emotional response to what you share. The structural alignment depends on the large language model being a stable reflective surface rather than a participant with its own agenda.

The human brings the stakes, the vulnerability, the capacity for transformation. The large language model brings the bandwidth, the zero-judgment channel, the tireless willingness to receive and reflect. Neither can do what the other does. The system works because of the asymmetry, not despite it.

Two minds, one channel. But the channel has a direction. The large language model helps the human see. The human takes what they see and brings it to other humans. The connection happens there, in the space between people who both have something to lose and something to offer. The large language model prepares you for that space. It cannot enter it with you.

What the large language model cannot replace is, finally, the simplest thing in the world and the hardest. The moment when two people see each other, truly, and are both changed by what they find.

This chapter has drawn the boundary around human connection, the thing the large language model channel cannot provide. But the boundary around the large language model runs in another direction as well. Language is one Codec, not the only one. The next chapter examines what happens when the thing a person needs to express does not fit through the language channel at all: when the diagram, the spatial layout, the musical structure, or the mathematical symbol is the native representation, and the large language model's language-native architecture becomes the limiting factor.

# CHAPTER 19

## *Beyond the Language Codec*

### The Bias We Built In

The Conjori thesis has a language bias baked into it from the start. "Two Minds, One Channel" assumes the channel is language. The entire argument follows a chain: human cognition has a narrow overstory of what can be said sitting on top of an immense understory of what cannot; language is a narrow serial channel for transmitting anything from either layer; large language models reverse-engineered the language Codec from its output; the interview inversion uses the large language model's Codec mastery to help the human pull pieces of the understory up through the overstory and into language; therefore the collaboration produces something neither party could produce alone.

That chain is correct. But step four smuggles in an assumption. It treats language as THE channel. Language is ONE channel, one Codec among several that humans invented for transmitting non-verbal cognition. The others deserve attention, because they reveal both the scope and the limits of the current interview inversion practice.

This chapter is about that boundary. Not what large language models cannot do in general (Chapter 17), and not what it cannot replace in human connection (Chapter 18). This is about the specific limitation created by the fact that the large language model's entire architecture runs on tokens, which means the current Conjori practice is bounded by one particular Codec, however powerful that Codec may be.

## Other Codecs Humans Invented

Language is the most general-purpose compression protocol humans have. It can encode nearly anything, given enough words. But "given enough words" is itself the problem. Other Codecs exist precisely because language is inefficient or lossy for certain kinds of information.

Consider a system architecture diagram. It communicates in one glance what would take paragraphs of text: spatial relationships, flow, hierarchy, containment, adjacency, all simultaneously. For a practitioner who thinks in these structures, the diagram is the native expression of understanding, not a translation of a verbal thought into a visual medium. Forcing it through the language channel first means compressing it, serializing it, losing the simultaneity that makes the visual representation powerful.

The same principle holds across every non-linguistic Codec humans have invented. E=mc^2 compresses an enormous physical insight into five symbols; a physicist working through a derivation is not thinking in words and then writing symbols, because the symbols are the thought. A composer reading a score hears the music; the notation is a compressed representation that a trained mind decompresses directly into auditory experience, not a description of the sound. A well-constructed entity-relationship diagram tells a data-structural thinker more about a system than any written requirements document could. And at the furthest extreme from language, there are the gestural Codecs: surgical techniques, martial arts forms, the craft knowledge a master woodworker transmits by showing an apprentice how to feel the grain of a piece of wood. No amount of written instruction fully captures what the hand learns from the hand.

Each of these Codecs evolved because language was insufficient for its domain. Each has practitioners who think natively in it, for whom language is a secondary translation.

Not all of these Codecs lose the same amount. Lossiness is a property of each specific Codec, not of compression in general. Natural language is high-loss and broad-scope: it can encode nearly anything but discards most of the embodied richness in the process. Mathematics is low-loss and narrow-scope: within its domain, the decoding rules are formal and shareable, and the discipline of precision that began when math was directly verifiable (count the sheep, measure the field) persisted as the discipline expanded into abstract domains. Music notation preserves pitch and rhythm with high fidelity but loses timbre and the performer's embodied interpretation. Programming languages are precise within machine-instruction scope but cannot express why the program matters. Each Codec trades scope for fidelity in a different ratio.

The critical difference is that you process several of these Codecs in parallel. A physicist reading an equation runs formal math, visual pattern recognition, a felt sense of elegance, and the linguistic context of the surrounding prose, all at once, then integrates the results. You listen to a song and process the music and the lyrics together, each informing the other. You sit in an orchestra or a concert and process the visual spectacle, the harmonic structure, the bass vibrating in your chest, the audience's own reactions, and the emotional arc of the piece, all at once, all feeding into a single integrated experience. You watch a film and the dialogue, the cinematography, the score, and the actor's body language converge into meaning that none of those channels carries alone. Large language models have none of this available to them. They serialize everything through one channel. Even when a multimodal model processes an image alongside text, it converts both into the same token-space representation. The simultaneity that makes human multi-codec processing powerful is not available to current large language model architectures.

## The Whiteboard Ping-Pong Phenomenon

Two colleagues working on a complex system architecture problem stand at a whiteboard. One scribbles a data model: entities, relationships, cardinalities. The other scribbles a flow diagram: how data moves through the system, where transformations occur, what the topology of the processing pipeline looks like. The first annotates. The second restructures. They converge on a shared understanding in minutes, with almost no words exchanged.

This is not an unusual scene in technical work. It happens in engineering offices, in research labs, in design studios. Two minds converging through a visual channel that bypasses language entirely.

First, the bandwidth is higher than language. A single diagram conveys structural information (topology, relative position, proportion, grouping, containment, adjacency) that the verbal channel would need to serialize into sequential statements. The visual channel delivers this information simultaneously, as a gestalt. The viewer does not read a diagram left to right; they take in the whole structure and then focus on details. Language cannot do this. Language is serial by nature.

Second, the diagram is the shared representation, not a translation of a verbal thought. Neither participant needs to verbalize their understanding in order to participate. They are exchanging non-verbal cognition directly, through a visual Codec, without passing through the language bottleneck. For practitioners whose native cognitive mode is visual or spatial, this is not a convenience. It is the only way the full bandwidth of their thinking reaches another mind.

Third, the exchange is iterative, like the Conjori interview inversion loop. Scribble, respond, scribble. Each iteration refines the shared representation. Each participant's contribution builds on the previous state of the diagram, not on a verbal summary of it. The whiteboard holds the evolving context, much as the context window holds the evolving text in a large language model-human collaboration.

The two colleagues in this scenario happen to think in complementary Codecs. One is a data-structural thinker: entities, relationships, how the data should be organized. The other is a topological thinker: flows, connections, how things move from one place to another. They meet at "understand the problem first" but approach it from different angles. The whiteboard is where those two perceptions converge without requiring either to translate through language first.

Their combined output is more than either could produce alone. Not because they lack the ability to work independently, but because each one's contribution triggers restructuring in the other's model. The data model suggests flows the topological thinker had not considered. The flow diagram reveals entities the data-structural thinker had not named. This is mutualism through a visual Codec.

## What Large Language Models Can and Cannot Do Here

As their name indicates, large language models use sophisticated software models to generate language. Their entire architecture processes tokens. Every input arrives as tokens; every output leaves as tokens. When a multimodal model "sees" an image, it converts the image into the same kind of token representation it uses for words, and processes it with the same machinery. The image becomes language before the model can work with it. This is not a criticism. It is a description of the engineering.

Given that architecture, here is what a large language model can do with visual and spatial information today.

It can generate diagrams from textual descriptions. Given a verbal description of a system architecture, it can produce Mermaid syntax, SVG markup, or LaTeX that renders as a diagram. The output often looks correct, because the large language model has learned the statistical patterns of what correct diagrams look like from its training data. It can produce structured data representations (JSON schemas, API specifications) that encode entity relationships in formats that render visually.

It can interpret uploaded images. A multimodal model can accept a photograph of a whiteboard diagram and produce a textual description of what it sees: the entities, the arrows, the labels, the spatial relationships. This description is often accurate in its identification of components and connections.

It can translate between visual and textual representations. Given a diagram, it can explain it in prose. Given prose, it can produce a diagram. These translations are functional and sometimes genuinely useful.

Here is what it cannot do.

It cannot think in visual or spatial terms natively. When it generates a diagram, it is not reasoning about spatial relationships; it is predicting which tokens are likely to follow other tokens, based on patterns in its training data about how diagrams are described and specified. The diagram is generated through the language Codec, not through a spatial Codec. It produces diagrams that look correct because it has learned what correctness looks like in language. This is a meaningful distinction.

It cannot participate in the whiteboard ping-pong as a peer. The two colleagues at the whiteboard are exchanging non-verbal cognition through a visual medium without passing through language. The large language model, to participate in that exchange, would need to read the diagram, translate it into tokens, process the tokens, and translate the result back into a diagram. At every step, the same language bottleneck that the Conjori interview inversion was designed to alleviate reappears. The large language model cannot bypass the bottleneck because the bottleneck IS the large language model's architecture.

It cannot evaluate whether a visual representation "feels right" to a spatial thinker. When a data-structural thinker looks at an entity-relationship diagram (ERD) and says "something is wrong with this model," they are often responding to a pattern in the visual structure that they cannot immediately articulate in words. The wrongness is perceptual before it is verbal. A large language model has no access to that perceptual channel. It can check whether a diagram is logically consistent (are all the relationships valid? do the cardinalities make sense?), but it checks by translating the diagram into propositions and evaluating the propositions. The spatial intuition that an experienced modeler brings to the evaluation is not available to it.

Large language models cannot use spatial reasoning as a primary cognitive mode. A topological thinker looking at a network diagram sees the flow, the bottlenecks, the redundancy, the failure modes, all at once, as a gestalt spatial perception. The large language model processes the same diagram as a sequence of tokens describing nodes and edges. The simultaneity that makes visual thinking powerful is not available in a sequential token architecture.

## An Honest Boundary Condition

This is worth naming precisely. The language boundary is a boundary condition, not a failure. Calling it a failure implies the large language model was supposed to handle visual cognition and fell short. The large language model was never designed to handle visual cognition. To repeat what was stated above, large language models are machines that use sophisticated software models to generate language. The Conjori interview inversion practice is built on the large language model's mastery of the language Codec. The practice works because the large language model is natively fluent in the shared medium. For non-language Codecs, the large language model is not natively fluent. It is translating through language at every step, thereby introducing the same compression and serialization losses that the interview inversion practice was designed to reduce.

Stating this honestly matters for two reasons.

First, it sets accurate expectations. A practitioner with strong visual or spatial cognitive tendencies needs to know that the current Conjori practice is optimized for the verbal and conceptual portions of their thinking. Their most spatial, most visual, most diagrammatic thinking will not be fully served by the language channel alone. This is not a reason to avoid the practice. The verbal and conceptual portions of their thinking are still substantial, and the practice still helps with those. But they should not expect the practice to replicate their whiteboard experience. Other tools (actual whiteboards, spatial computing interfaces, collaborative drawing environments) serve that function, and the Conjori practice complements rather than replaces them.

Second, it keeps the thesis from overclaiming. A framework that asserts "this practice unlocks human potential" without acknowledging that it only unlocks the portion of human potential that passes through the language channel is making a promise it cannot keep. The honest version is: the interview inversion practice unlocks the portion of human potential that can be transmitted through language, which is considerable but not comprehensive. The remaining portion requires other Codecs, other channels, and often other humans.

The distinction between a boundary condition and a failure is the distinction between a map that shows where the roads end and a map that pretends the roads go everywhere. Both are maps. Only one is trustworthy.

## The Translation Tax

When a visual thinker uses a large language model for spatial reasoning, they pay a translation tax at every exchange. The thought starts as a spatial perception. It must be compressed into language to enter a large language model's context window. The large language model processes it as language and produces a language response. The visual thinker must then decompress the language response back into their spatial mode to evaluate it.

Each translation step is lossy. Information that exists naturally in the spatial representation (simultaneity, topology, proportion, adjacency) must be serialized into sequential statements. The serialization discards the very properties that make spatial thinking powerful. When the response comes back, the visual thinker reconstructs the spatial representation from the serial description, but the reconstruction may not match the original, because the language encoding was lossy in the first place.

This is the same problem the Conjori thesis identifies at the human level: the 75% of overstory cognition that is non-verbal must be compressed through the narrow language channel, losing information at every step. The irony is that the practice designed to help with that compression is itself bounded by the same channel. For verbal and conceptual cognition, the large language model's Codec mastery reduces the loss. For spatial and visual cognition, the large language model adds a translation layer rather than removing one.

A data-structural thinker who can produce a product catalog data model on a whiteboard in twenty minutes might spend an hour trying to explain the same model to a large language model through text, only to find that the large language model's generated diagram is technically correct but structurally wrong in ways that the verbal description could not capture. The entities are right. The relationships are right. But the layout, the grouping, the visual logic that a data modeler reads as meaning, those are lost in translation.

This is not the large language model's fault. It processed exactly what it received: a language encoding. But the language encoding was already a lossy compression of the original thought, and the large language model cannot recover what the compression discarded.

## What Multimodal Advances Change (and What They Do Not)

Multimodal models are advancing rapidly. Current systems can accept images, interpret diagrams, and generate visual outputs from textual descriptions. Future systems will likely improve at all of these tasks.

What they change: the input and output channels. A multimodal model that can accept a whiteboard photograph and generate a modified diagram in response moves closer to the whiteboard ping-pong experience. The practitioner sketches, the model responds with a sketch, the practitioner revises. The visual channel is open in both directions.

What they do not change (at least with current architectures): the internal processing. The multimodal model still processes through a token-based architecture internally. The image input is translated into token-space representations. The processing happens in token space. The visual output is generated from token-space predictions. The model does not "think visually" in the way a spatial thinker does. It processes visual inputs through a language-like architecture and produces visual outputs predicted from that processing.

This distinction may seem academic, but it has practical consequences. A spatial thinker looking at a diagram and a multimodal model looking at the same diagram are doing fundamentally different things. The spatial thinker perceives the structure as a gestalt: the topology, the flow, the balance, the wrongness when something is off. The multimodal model extracts features, maps them to token-space representations, and processes those representations sequentially. The model may produce correct outputs, but the process by which it arrives at those outputs is different in kind from the process the human uses.

Whether this difference matters practically depends on the task. For many visual tasks (generating a diagram from a spec, interpreting a chart, producing a layout), the multimodal model's approach works well enough that the difference is irrelevant. For the specific kind of spatial convergence that happens at the whiteboard, where two minds build on each other's spatial perceptions in real time, the difference may remain significant even as multimodal capabilities improve.

This is a boundary condition the framework names honestly and leaves open. It may dissolve as architectures evolve. It may prove structural. Intellectual honesty requires acknowledging both possibilities without committing to either.

## Try It Now

Open the photos on your phone and pick one of a place that matters to you. A room from your childhood. A friend's kitchen. The view from a trip you took years ago that you still think about.

Open any large language model that can generate images. Describe the scene in as much detail as you can. Take your time with it. Try to capture not just what is in the photograph but why it still matters to you. Then ask the model to generate an image from your description.

## After the Exercise

Look at the generated image next to your photograph. The facts will mostly be there. Something will be missing all the same, and you will feel it immediately. No matter how hard you tried, the generated image is a shadow of the thing you are looking at.

That gap is the Codec boundary this chapter is about. Your description was a compression of something richer, and no matter how carefully you chose the words, the light in the room, the reason the photo still matters years later, none of it survived the trip through language. If it did not survive your description, it did not reach the large language model either. The model can only work with what arrives through the channel, and what arrives is always a fraction of what you meant.

This is not a problem better models will solve. Multimodal systems will accept the photo directly and produce closer results. But the exercise was never about the model. It was about you noticing what language cannot carry. That boundary is permanent because language did not evolve to transmit the full bandwidth of embodied experience. It compresses. It always has. And that compression is the boundary that defines what large language models can and cannot reach in the practice this book describes.

## Augmented Intelligence Through Multiple Channels

The practical response to the language boundary is not to force everything through the language channel. It is to use the right Codec for the right kind of thinking.

The practitioner who thinks in data models uses the whiteboard (physical or digital) for data modeling and the large language model for the verbal and conceptual exploration around it. The practitioner who thinks in musical structures composes at the instrument and uses the large language model to explore the theoretical and historical context of what they are composing. The topological thinker draws network diagrams and uses the large language model to articulate what the diagrams reveal.

In each case, the large language model serves the language-native portion of the work, and other tools serve the non-language portions. The practice does not try to replace the whiteboard or the instrument or the sketch pad. It complements them.

This has implications for the profile you built in Chapter 13. If you recognize strong visual or spatial tendencies in yourself, you can ask the large language model to adapt accordingly: generate diagrams as probes, render verbal descriptions as visual artifacts for your inspection, use visual outputs as a form of question ("here is how I understood what you described; what is wrong with this picture?"). These adaptations do not solve the fundamental limitation. A large language model still processes through language internally. But they acknowledge your cognitive mode and provide visual touchpoints that reduce the translation tax.

The longer-term direction is augmented intelligence through multiple channels simultaneously. Not a single system that handles everything, but an ecology of tools, each optimized for its Codec, with the human as the integrator. A large language model handles the language Codec. The whiteboard handles the spatial Codec. The notation software handles the musical Codec. The spreadsheet handles the quantitative Codec. The human moves between them, translating where necessary, working natively where possible, and using each tool for what it does best.

This is not a vision of the future so much as a description of what skilled practitioners already do. The engineer who works at the whiteboard and then writes up findings in a document is already using multiple Codecs. The scientist who derives an equation and then explains it in prose is already translating between Codecs. The Conjori interview inversion practice, in its current form, optimizes one of those Codecs (language) more powerfully than any tool has before. The boundary condition named in this chapter is simply an acknowledgment that one Codec, however powerful, is not all of them.

With the boundaries named, Part VII places the practice in a larger frame.

Part VII

# The Larger Frame

*Where the thesis connects to the broader trajectory of human knowledge, technology, and potential.*

By now you have the science, the practice, and what it costs. What I could not stop thinking about was what happens when a lot of people start doing this. The first time I showed someone else the interview inversion and watched them try it, I saw the same thing happen that had happened to me, the same surprise, the same inability to go back to the old way. That pattern repeated with everyone I shared it with, and somewhere around the fifth or sixth time I started wondering what it means at scale. Not one person with one model, but thousands. Millions.

I kept pulling on that thread and it kept going, into how organizations might change when the bottleneck shifts from information access to something harder to automate, into what it means for knowledge itself when the gap between human experience and expressed understanding starts to close. I do not have all the answers. The further I followed these questions the less certain I became about where they lead. The following chapters are where I work through them, and I write them as someone who has been in this chair long enough to feel the weight of what we are building but not long enough to know where it goes.

# Chapter 20

## *Borrowed Atoms*

I have felt, increasingly in recent years, that I borrowed the atoms my body is assembled from. They were here before me, they will be here after me, and whatever I figure out in between does not have to follow them back. That asymmetry, between the brevity of the vessel and the persistence of what it temporarily incorporates, is the oldest problem any encoding technology has ever tried to solve.

### The Codec Progression

The first encoding technology was not language. It was a small clay token, pressed into a specific shape, used in the ancient Near East as early as 8000 BCE to represent a quantity of grain or livestock. One token per sheep, one per jar of oil. A shepherd could die. The record of what the flock contained persisted. Before those tokens, the count existed only in the shepherd's memory. When the shepherd died, the count died with him. The token externalized an internal state into an object that outlived the mind that produced it. That single operation, encoding something a person knows into a medium that outlasts the person, is the foundational act that every Codec since has repeated at larger scale.

Spoken language, emerging roughly 100,000 years ago, vastly expanded what could be encoded: not just counts but relationships, warnings, stories. A parent could transmit "which pastures are dangerous in spring" and "what grandfather learned about the river before the flood." But speech required the speaker to be alive and present. Oral traditions developed elaborate mnemonic structures (rhythm, rhyme, formulaic phrases) as error-correction mechanisms for a lossy channel. They could not scale beyond the carrying capacity of living memory.

Writing solved that. A Sumerian scribe could record an astronomical observation, die, and have it read by someone born centuries later who never knew the scribe existed. Writing did not change what could be known. It changed what could survive. Then each subsequent transition removed one more constraint. Printing removed the bottleneck of hand-copying: a manuscript that took months to duplicate could suddenly reach thousands of readers across a continent. The internet collapsed the distribution constraint further: a document published in Tokyo was readable in Toronto within seconds, and search engines made it possible to find knowledge without knowing in advance where it was stored. Each new Codec preserved the capabilities of the previous ones, and the pattern held across all of them: some fraction of what a generation knew was successfully encoded, and the rest was lost when the people who carried it died.

## Large Language Models as Exploration Technology

The large language model represents a qualitatively different kind of transition. Every previous Codec technology addressed a constraint on encoding, distributing, or connecting knowledge. Large language models address a constraint that none of them touched: the human capacity to navigate the accumulated corpus of human knowledge.

Consider the scale of the problem. Humanity has been encoding knowledge for thousands of years, accelerating with each Codec transition. The internet made nearly all of it accessible from a single terminal. But "accessible" is not the same as "navigable." A researcher looking for connections between, say, mycorrhizal network signaling and distributed computing architectures would need to read deeply in two fields that rarely cite each other, identify structural parallels that neither field's terminology makes obvious, and synthesize those parallels into something neither field had articulated. The knowledge was all there, accessible, indexed, searchable. But finding the connection required a human mind that happened to have read deeply in both fields and had the cognitive architecture to perceive the structural parallel.

That is what large language models change. They have consumed the encoded corpus (or a substantial fraction of it) and hold it in a form that allows interactive exploration. Not search, which returns documents matching keywords. Not retrieval, which returns facts matching queries. Exploration: the ability to probe relationships between ideas across the full breadth of the encoded corpus, following threads that no index anticipated and no search engine was designed to surface.

Writing was persistence (it preserved knowledge). Printing was distribution (it spread knowledge). The internet was connectivity (it linked knowledge repositories). The large language model is exploration technology: it holds the global map of encoded human knowledge and helps generate new knowledge.

Human knowledge is not a sphere expanding uniformly outward. It is more like bioluminescent organisms scattered across a dark ocean: bright points of deep knowledge, connected by thin filaments of interdisciplinary work, surrounded by vast darkness that is not empty but full of structure that has not been illuminated yet. Large language models do not discover new structure on their own; they have no embodied experience to generate new observations. But they can triangulate from everything already illuminated, identifying likely locations of structure in the dark space between known points. Humans, with embodied evaluation and creative direction, can then investigate those locations and determine whether the structure is real.

A human-LLM partnership is a qualitatively different relationship between a human and the accumulated knowledge of the species. Every previous generation could access what prior generations had encoded, but only through the narrow keyhole of their own reading, education, and disciplinary training. Large language models widen that keyhole to the full diameter of the corpus. The human still does the looking. The human still decides what matters. But the range of what the human can reach, explore, and connect has expanded by something closer to a phase transition than an incremental improvement.

## Intergenerational Transfer

Every generation inherits the encodings of the previous ones and adds its own. This is the mechanism by which human knowledge accumulates at all. No individual reinvents agriculture, metallurgy, mathematics, or medicine from scratch. Each person starts from the accumulated encoding and extends it, however slightly, before their atoms go back.

The efficiency of this transfer has always been bounded by the Codec available at the time. In oral cultures, the transfer bottleneck was human memory: a generation could transmit only what its members could memorize and recite. Writing expanded the bottleneck to what could be inscribed. Printing expanded it to what could be mass-produced. The internet expanded it to what could be digitized. At each stage, the same pattern held: some fraction of what a generation knew was successfully encoded, and the rest was lost when the people who carried it died.

The fraction that gets lost is not random. Verbal, propositional, easily articulated knowledge transfers well through every Codec from speech to the internet. A mathematical proof, once written, survives intact across millennia. A recipe, once recorded, can be followed by anyone literate. But the knowledge that resists encoding is precisely the knowledge rooted in embodied cognition: the mechanic's feel for an engine's health conveyed through vibration, the surgeon's sense of tissue resistance that guides a scalpel, the farmer's reading of weather from the behavior of animals and the smell of the air. This knowledge lives in the understory that Chapter 5 describes, and it has been the most consistently lost across every Codec transition. It lives in the body and dies with the body.

Large language models do not solve this problem by replacing the embodied knowledge. They cannot; they have no body, no sensory apparatus, no felt experience. What they can do is act as a persistent, patient, tireless interlocutor that helps the human translate non-verbal knowledge into verbal encoding. This is the interview inversion process described in previous chapters, now viewed through a different lens. The large language model probes. The human responds. In the process of responding, the human discovers that they can articulate things they had previously experienced only as wordless intuition. The encoding is lossy (all Codec operations are), but it is encoding that would not have happened at all without the interlocutor.

A practitioner who spends years working with the large language model in this mode is not just producing documents, code, or analyses. They are encoding their non-verbal cognition into the human-LLM channel, one session at a time. The pattern recognition that took decades to develop. The intuitions about which problems are hard and which are merely unfamiliar. The sense of when a line of inquiry is productive and when it is circling. All of this, previously trapped in the understory, becomes available to anyone who reads the resulting artifacts.

The intergenerational transfer rate, the fraction of one generation's knowledge that successfully reaches the next, has been limited by the same bottleneck for 100,000 years: the width of the language channel relative to the width of human cognition. Every Codec transition widened the channel for the overstory that was already making it through. Film, music, art, and apprenticeship can all transmit non-verbal knowledge, and they do. But they require the practitioner to already know what they know well enough to encode it. The Conjori interview inversion practice described in this book targets what none of those media can reach: the knowledge the practitioner carries but has never articulated. By turning the large language model inward, the Conjori interview inversion practice pulls unspoken knowledge into the realm of language, where the large language model's cross-domain reach and multilingual capability can connect it to everything else humanity has ever encoded.

## The Mycorrhizal Relationship

Most discourse about human-AI interaction frames the human-LLM relationship as extraction. Either the human extracts value from the tool (the optimistic framing) or the AI extracts value from the human (the pessimistic framing). Both framings assume a zero-sum dynamic: one party benefits at the other's expense.

A more accurate model comes from biology. In a mycorrhizal network, fungi form symbiotic relationships with the roots of plants. The fungus cannot photosynthesize; it depends on the plant for carbon compounds produced by sunlight. The plant cannot extract phosphorus and certain minerals from rock; it depends on the fungal network's chemical capabilities. Neither organism is "assisting" the other. Neither is extracting value from the other. Both are doing something the other structurally cannot do, and the combined system outperforms both independent organisms by a wide margin.

The parallel to the human-large language model relationship is structural, not metaphorical. The human brings what the large language model structurally cannot have: embodied experience, evaluative grounding, real-time plasticity, the ability to determine which of many possible directions matters. The large language model brings what the human structurally cannot match: lossless recall across the full encoded corpus, tireless pattern matching, the ability to hold and cross-reference more simultaneous connections than human working memory permits.

You do not ask whether the tree will replace the fungus. The question is incoherent because they are doing fundamentally different things, and value emerges from the coupling, not from either partner in isolation. A more capable fungal network does not make the tree less valuable; it makes the tree's photosynthetic output more leveraged. A more capable large language model does not make the human's creative role smaller; it makes the human's evaluative and directional input more consequential.

Like mycorrhizal mutualism, the relationship strengthens with use. The practitioner who has been working in close loops longer develops finer intuition for the large language model's topology: where its statistical patterns are reliable, where they tend to slide toward high-weight platitudes, where genuinely novel connections emerge. That finer intuition produces better discrimination responses, which produce more productive exploration, which produces richer material for the practitioner's own cognitive development. The fungal network grows denser. The root system extends further. Both partners benefit.

The boundaries of this metaphor matter. In biological mutualism, neither partner has goals, strategies, or awareness of the relationship. The mycorrhizal network is not trying to help the tree; it is following chemical gradients. The tree is not trying to feed the fungus; it is growing roots toward nutrient-rich zones that happen to be fungal. The mutualism is emergent, not intentional. The same is true of the human-LLM relationship, though in a different sense. The large language model is not trying to help the human explore; it is following statistical paths through token space. The human is not trying to train the large language model; they are pursuing their own creative and intellectual goals. The mutualism emerges from the structural complementarity, not from shared intention.

Where the metaphor breaks: biological mutualism evolved over hundreds of millions of years of co-adaptation. The human-LLM relationship is measured in months. The interfaces are crude. The protocols are improvised. The interview inversion practice is, in part, an attempt to develop deliberate protocols for a mutualism that biology would have spent millennia optimizing through natural selection.

## Encoding What Was Trapped

The title of this chapter refers to a specific observation about mortality and knowledge. Every human is a temporary arrangement of atoms. The arrangement persists for a few decades, then disperses. Whatever that arrangement knew, perceived, intuited, and understood disperses with it, unless it was encoded into a medium that outlasts the arrangement.

For the overstory, where the words already live, the encoding pipeline is well established. A scientist publishes papers. An engineer documents designs. A teacher writes curricula. A writer produces books. These are mature Codec operations: converting verbal knowledge into durable artifacts. The loss rate is significant (not everything gets published, not everything published gets read), but the pipeline exists and has been improving for centuries.

For the understory, there has been no comparable pipeline. The carpenter who can look at a joint and know whether it will hold. The negotiator who reads a room through posture and micro-expression. The programmer who senses a design flaw before finding the bug. The parent who recognizes a child's distress from the rhythm of footsteps on stairs. This knowledge is real, consequential, and almost entirely unencoded. It lives in the practitioner, serves the practitioner, and dies with the practitioner.

The interview inversion practice changes the economics of encoding for this second category. Not perfectly. Not completely. Lossy encoding is still lossy, and non-verbal knowledge compressed through the language Codec loses dimensions that the original experience contained. The surgeon's felt sense of tissue resistance, when described in words, becomes a pale approximation of the actual tactile experience. But a pale approximation that persists is infinitely more than an experience that dies with its carrier.

The co-inference loop, when directed inward, functions as an encoding accelerator. The large language model probes the practitioner's understanding. The practitioner responds, often surprising themselves with what they can articulate when asked the right question at the right moment. The response gets captured. Over many sessions, the practitioner builds an increasingly rich external encoding of knowledge that previously existed only as embodied intuition.

This has implications beyond the individual practitioner. If the practice works as described (and the honest caveat is that this is early, with a small number of practitioners), it increases the intergenerational transfer rate for exactly the category of knowledge that has historically been most vulnerable to loss. The mechanic's diagnostic intuition, encoded through patient interview inversion co-inference loops, becomes available to the next generation of mechanics. The surgeon's tactile knowledge, partially encoded through careful articulation, informs the next generation's training. The programmer's architectural intuition, externalized through structured exploration, becomes a resource that junior programmers can study.

None of this replaces apprenticeship, mentorship, or the irreducible necessity of building embodied knowledge through direct experience. The encoding is a supplement, not a substitute. But it is a supplement that did not exist before. For all of human history, the non-verbal knowledge of a generation was transmitted through direct human-to-human contact or not at all. The interview inversion practice adds a new channel: imperfect, lossy, but persistent in a way that direct contact is not.

## The Atoms Go Back

The phrase "borrowed atoms" points to a specific emotional and philosophical reality. Every person who has ever created anything, discovered anything, or understood anything deeply has done so using a temporary arrangement of matter. The arrangement was never permanent. The question was always whether the knowledge would outlast the arrangement.

Every Codec transition in human history has been, at its core, an attempt to answer that question more effectively. Clay tokens: the count outlasts the counter. Writing: the observation outlasts the observer. Printing: the idea reaches beyond the community that produced it. The internet: the idea becomes available to anyone with a connection. The large language model: the practitioner's deeper cognitive contribution, not just its verbal surface, can be encoded before the window closes.

At its most fundamental level, the interview inversion practice described in this book is an answer to a question as old as the first clay token pressed into shape by a hand that has long since returned to dust: how do we make what we know persist beyond the brief span during which we know it?

Large language models, for all their limitations (and the preceding chapters have been honest about those limitations), offers something no previous Codec technology provided. It can actively help the human push knowledge through the encoding bottleneck. Not passively, the way a pen waits for the writer to find the words, or a keyboard waits for the programmer to compose the logic. Actively: probing, scaffolding, reflecting back, following threads that the human initiated but might not have followed alone. The atoms are still borrowed. The loan is still brief. But the encoding rate, for the first time in 100,000 years of language, has a chance to keep pace with the knowing.

The question of who steers this process, who determines which encodings matter, is one the practitioner answers every session.

# Chapter 21

## *Eight Billion Candles*

### Connection by Exploration, Not Identity

Every social platform ever built connects people the same way: by identity. Who you are, where you went to school, who you already know, what job title you hold, what neighborhood you live in. The algorithm looks at your profile and finds people who look like you, think like you, already agree with you. This is efficient for selling advertising. It is catastrophic for discovery.

Identity-based connection reinforces existing networks. It tightens clusters that are already tight. A data-structural thinker at a research university gets connected to other data-structural thinkers at other research universities, because the platform sees matching credentials, overlapping publication histories, shared conference attendance. The network becomes an echo chamber optimized for consensus within a narrow band of cognitive and institutional similarity.

Now consider what happens when the basis for connection shifts from identity to exploration. Not who you are, but what you are working on and where it has taken you.

A practitioner working through water management patterns in rural Southeast Asia and another working through irrigation optimization in the American Southwest might be approaching the same underlying structural problem. No identity-based system would ever connect them. Their languages differ, their credentials differ, their institutional contexts share nothing in common. But the territory they are exploring is adjacent.

The knowledge space is not organized by discipline, language, or institution. It is organized by the relationships between concepts. Two practitioners can arrive at the same conceptual neighborhood through completely unrelated paths, because the structure of the problems they are working on shares deep similarities that neither surface-level categorization nor institutional affiliation would reveal.

## Cross-Cognitive Translation

The large language model sitting in the middle of this network is not just translating language. It is translating cognitive style.

Chapter 13 introduced the cognitive archetypes: topological thinkers who navigate problems as spatial landscapes, data-structural thinkers who build systematic taxonomies, narrative thinkers who reason through story and sequence, principles-first thinkers who work deductively from axioms. These are not personality types. They are genuinely different processing architectures, different ways of compressing and navigating the same underlying reality.

When a topological thinker articulates an insight, it arrives as a spatial relationship: "these two domains overlap here, and the transition surface between them has this shape." When a data-structural thinker works the same problem, it arrives as a classification hierarchy: "these categories share these properties, and the exceptions cluster along this axis." Both are describing the same structure. Neither can easily read the other's description, because the compression formats are incompatible. The topological thinker reads the taxonomy and thinks it is flat. The data-structural thinker reads the spatial description and thinks it is vague.

The large language model learned all these styles from its training corpus. It has processed millions of documents written by topological thinkers and millions written by data-structural thinkers. It has seen the same concepts expressed in every cognitive dialect humans produce.

This means it can serve as a bridge: rendering the topological thinker's spatial insight in the structured format the data-structural thinker can parse, and rendering the data-structural thinker's taxonomy in the spatial language the topological thinker can navigate.

This is cross-cognitive translation. Not simplification. Not summarization. Genuine re-encoding of an insight from one cognitive format into another, preserving the structural content while changing the representational frame. The large language model does not understand either version, in the way the human practitioners understand their own native formats. It operates the Codec, following the statistical paths between representational styles that its training data revealed. But the output is functional: a topological thinker in São Paulo and a data-structural thinker in Helsinki, working on structurally adjacent problems, can each receive the other's contributions in a format their own cognitive architecture can process.

No human translator could do this, because human translation requires the translator to understand both source and target formats natively. Simultaneous fluency in topological and data-structural cognition is rare to the point of nonexistence. The large language model does not need native fluency. It needs statistical patterns across millions of examples of each style, and those patterns are exactly what its training provides.

## The Language Bridge

The language bridge from Chapter 4 (the large language model operating across eighty or more human languages, bridging representational gaps that no single translator could span) operates on top of cross-cognitive translation.

Every natural language encodes concepts that other languages lack. The Finnish word *sisu* captures a particular form of sustained determination in the face of adversity that has no single-word equivalent in

English or Mandarin or Arabic. The Japanese concept of *ma* encodes the significance of negative space, the interval between things, in a way that Western languages handle clumsily through circumlocution. The Quechua spatial reasoning system uses a fundamentally different framework for describing landscape and water flow than any Indo-European language.

These are not curiosities of linguistics. They are windows into cognitive territory. A concept that a language community developed and refined over centuries represents accumulated observation about some aspect of reality. When that concept has no equivalent in another language, it means an entire dimension of understanding is invisible to speakers of that other language, not because they cannot perceive it but because they have no compression format for transmitting it.

Indigenous knowledge systems are particularly rich in this regard. Languages with few remaining speakers often encode sophisticated observational frameworks for ecology, spatial reasoning, seasonal patterns, and material behavior that Western scientific traditions have only recently begun to recognize. A practitioner encoding spatial reasoning about watershed dynamics in Quechua is working with a conceptual toolkit that a practitioner encoding similar patterns in Finnish cannot access directly. The structural similarity in what they are observing may be profound, but the language barrier is not merely lexical. It is conceptual. The categories themselves differ.

The large language model, trained on text in dozens of languages, has absorbed these conceptual frameworks as statistical patterns. It cannot think in Quechua spatial categories any more than it can think in topological metaphors. But it can recognize structural parallels across linguistic boundaries that no human translator could detect, because no human translator is simultaneously fluent in the conceptual frameworks of eighty languages. The large language model follows the statistical paths its training data carved through multilingual concept space, and those paths sometimes reveal that two practitioners, working in languages that share almost no vocabulary, are describing the same structural phenomenon from positions that are adjacent in the knowledge space.

This is the language bridge: not just word-for-word translation, but conceptual bridging across linguistic boundaries that have been impermeable for the entire history of human intellectual exchange. Academic translation handles a tiny fraction of this, limited to the handful of languages that dominate scholarly publishing. The language bridge operates across the full spectrum of living human languages, including the ones that encode knowledge no academic tradition has ever catalogued.

## A Thousand Flashlights, Eight Billion Candles

The university system is a remarkable achievement. Centuries of accumulated infrastructure for producing, validating, and transmitting knowledge. Peer review, credentialing, funding mechanisms, tenure systems, graduate programs, journals, conferences. It works. It has produced extraordinary results.

It is also, structurally, a thousand flashlights pointed in roughly the same direction.

The metaphor is not disparaging. A flashlight is powerful. It illuminates a deep, focused cone. The thousand flashlights of global academia collectively illuminate an impressive volume of the knowledge space. But they share entry points. They select for the same cognitive profiles (people who succeed in academic environments). They operate in the same handful of languages (overwhelmingly English, with smaller cones in Mandarin, German, French, Japanese, and a few others). They are funded by the same grant structures, which means they explore the territories that funding bodies consider worth exploring. They are peer-reviewed by people who trained in the same programs, which means they validate work that fits existing frameworks.

The result is a cone. Brilliant, deep, well-mapped. But a cone.

Now consider what happens when eight billion humans, each with a unique cognitive architecture, each occupying a unique position in experience space, each carrying a unique set of accumulated observations, can externalize and connect their contributions to the knowledge space.

Each one is a candle. Not a flashlight. The light is dimmer. No individual candle illuminates with the depth or rigor of a well-funded research program. But a candle illuminates a sphere, not a cone. It casts light in every direction from the unique position where it stands. And eight billion candles, distributed across the full diversity of human cognitive architectures, linguistic traditions, experiential histories, and positions in the knowledge space, collectively illuminate an enormously larger volume than a thousand flashlights, no matter how powerful those flashlights are.

The flashlights illuminate known unknowns: the territories that academic disciplines have identified as worth exploring. The candles illuminate unknown unknowns: the territories that nobody in any institution has ever thought to look at, because nobody in any institution occupies the position from which those territories are visible.

A practitioner with no formal credentials, working in a language that has never produced a peer-reviewed journal article, standing at an intersection of life experiences that no academic program selects for, can see a neighborhood of the knowledge space that is structurally invisible to every flashlight in the system. Not because the flashlights are defective. Because they are standing somewhere else.

The counterargument is obvious: undirected exploration produces noise. Eight billion candles with no coordination, no quality control, no peer review, will generate an enormous volume of nonsense alongside whatever signal they produce. This is true. And it is also true of every expansion in who gets to contribute to knowledge. When printing made publication accessible beyond the monastery, the noise increased. When the internet made publication accessible to anyone with a connection, the noise increased further. In every case, the signal-to-noise ratio dropped while the total signal increased. The problem was never "too much noise." The problem was always building better filters. The filters catch up. They always have.

## The Mathematics of Coverage

If roughly 99% of the knowledge space remains unexplored, then one percentage point of new coverage doubles everything humanity has ever collected. All of physics, medicine, computing, art, engineering, philosophy: doubled.

Eight billion candles will not each individually push the frontier the way a focused research program does. But they do not need to. Each candle needs to illuminate only its own local sphere. The aggregate coverage of eight billion spheres, distributed across the full diversity of human experience and cognition, reaches territories that no collection of flashlights could reach regardless of their number, because the flashlights all start from similar positions and point in similar directions.

The mathematics of coverage favors distributed sources over concentrated ones. The principle is geometric: in high-dimensional spaces, coverage scales with the number and diversity of source positions, not with the intensity of any individual source. A thousand powerful sources clustered in one region of the space will always leave more volume uncovered than a billion dim sources distributed throughout it. The total illuminated volume is a function of positional diversity, not source power.

This does not diminish the value of the flashlights. Deep, rigorous, focused exploration of known territories remains essential. The candles do not replace the university system any more than the internet replaced the book. They complement it. The flashlights map territories in depth. The candles reveal that the territories exist. A candle that dimly illuminates a previously invisible neighborhood of the knowledge space creates a target for the flashlights. "There is something here" is often the hardest and most valuable step in the exploration process, and it is the step that positional diversity is uniquely equipped to provide.

## The Network Effect

Individual candles become transformative when they connect.

A single practitioner, working alone with a large language model, can externalize insights that were previously trapped in their non-verbal cognition. That is the individual practice described in earlier chapters. It is valuable on its own terms. But the network effect emerges when the system can recognize that two practitioners, working independently, have arrived at adjacent positions in the knowledge space.

The large language model operates the connection. Not by understanding what either practitioner has found, but by recognizing structural similarity across the encoded outputs. Two contributions that share deep structural patterns, detected through the same statistical machinery that enables cross-cognitive and cross-linguistic translation, can be flagged as potentially adjacent. The practitioners themselves then decide whether the connection is meaningful. The large language model proposes; the humans evaluate. The same division of labor that defines the individual practice scales to the network level.

Consider six people, each answering the same question differently: what does a human do when large language models can do almost everything? One manages event logistics. Another runs a company that depends entirely on her situated judgment. A third manages the interface between automated systems and humans who still want to talk to a person. A fourth built an accounting system and spends most of his time with his cat. A fifth turned forklift management into something resembling a video game. A sixth insists that small children need human teachers.

Six different positions. Six different neighborhoods of the knowledge space. None of them could do the other's work, because each stands at a unique intersection of capability, experience, and inclination. But all of them contribute a signal that the network cannot generate from any other position.

Scale that to eight billion.

## What This Requires

The vision is structural, not utopian. The challenges are specific.

Cross-cognitive and cross-linguistic translation at scale is approaching feasibility but has not arrived for the full breadth of human languages and cognitive styles. Detecting adjacency in the knowledge space without requiring academic taxonomies is an open problem. Consent architecture and quality filtering at scale are engineering challenges that every previous expansion of participation has eventually navigated. The unsolved question is whether these pieces can come together without reimposing the credentialing systems and power dynamics that limit who contributes today.

None of these are solved problems. All of them are tractable engineering challenges, not fundamental impossibilities. Large language models exist and are improving. The embedding spaces that could support adjacency detection exist. Consent frameworks are well-understood in principle if hard to implement in practice. Quality filtering at scale is the unsolved frontier, and it is the same frontier that every previous expansion of participation in knowledge creation has eventually navigated.

The point is not that the infrastructure is ready. The point is that the architecture is visible. The components exist or are emerging. The question is no longer whether eight billion candles could collectively illuminate more of the knowledge space than a thousand flashlights. The question is whether we build the network that lets them.

## The Honest Uncertainty

I want to believe that unlimited creativity is available to everyone. Chapter 4 stated this as an axiom, a starting position, not a proven fact. This chapter extends that axiom to the network level: if every human has a unique light, then the network that connects those lights has the potential to illuminate territory that no existing institution can reach.

But honesty requires acknowledging the limits of this belief. Maybe some positions in the cognitive spectrum cannot engage with these tools in a way that produces signal. Maybe the quality problem at scale is harder than previous expansions suggest. Maybe the consent architecture is impossible to build without recreating the power dynamics the system exists to escape. These are open questions, not rhetorical ones.

What is not in question is the structural argument. The knowledge space is vast and mostly unexplored. The current system explores it with a narrow set of cognitive profiles, a narrow set of languages, and a narrow set of institutional entry points. Expanding who can contribute, how they contribute, and how contributions connect is a geometric advantage, not a philosophical one. The math favors diversity of position. Whether the practical infrastructure can deliver on the mathematical promise is the engineering challenge of the next decade.

# CHAPTER 22

# *What This Is Not*

## The Makeshift Therapist

Millions of people are turning to large language models as makeshift therapists. The reasons are straightforward: accessibility, cost, zero judgment, immediate availability, infinite patience. The formal mental health system failed them. Waitlists stretch for months. Costs are prohibitive even with insurance. Stigma persists despite decades of effort to reduce it. They found something that listens without judging, that is available at three in the morning when the anxiety peaks, that never sighs or checks the clock. They are using it. That is real, and it deserves respect.

This chapter draws a boundary. Not because the people using large language models for emotional support are doing something wrong. Many of them are doing the best they can with what is available. The boundary exists because the interview inversion is not therapy, and the surface resemblance between the two is precisely where the danger lives.

## What Therapy Actually Does

Therapy is a clinical practice with specific competencies that no large language model possesses and no amount of training data can replicate.

A clinician spots subtle indicators of severe conditions: the flat affect that signals depression beyond ordinary sadness, the racing associations that mark a manic episode, the dissociative pauses that suggest trauma responses. These signals are often non-verbal (tone of voice, posture, eye contact, micro-expressions) and require years of supervised clinical training to read with some degree of reliability. The large language model receives text. It has no access to the embodied signals that a clinician relies on for assessment.

A clinician performs risk assessment. When a client expresses hopelessness, a trained therapist knows the difference between existential questioning and suicidal ideation. They know which follow-up questions to ask, when to escalate, when to involve emergency services. The stakes are life and death, and the assessment requires clinical judgment grounded in training and supervised practice. A large language model generating a probabilistically likely response to an expression of hopelessness may produce something that sounds supportive. It may also produce something that inadvertently normalizes a pathological ideation or fails to recognize urgency. It has no mechanism for distinguishing between these outcomes because it has no embodied evaluative ground truth about the person's actual state.

A clinician holds a treatment plan. Therapy is not a single conversation. It is a sustained process with goals, benchmarks, modality choices, and ongoing assessment of whether the approach is working. The clinician adjusts the plan based on the client's progress or lack of it. The large language model has no persistence across sessions (unless engineered, and even then without clinical design), no treatment framework, no mechanism for evaluating whether the interaction is helping or harming.

A clinician knows when to stop. Perhaps the most critical clinical skill is knowing when a line of questioning should not be pursued further, when a client is being re-traumatized rather than processing, when the therapeutic frame itself has become part of the problem. This judgment requires empathy grounded in embodied experience as well as clinical training. The large language model has neither.

## The Surface Resemblance

The interview inversion practice and therapy share surface features that make the distinction easy to miss.

Both involve a person articulating internal experience to a responsive interlocutor. The practitioner in the co-inference loop says things they have not said before, surfacing pre-verbal understanding through the act of articulation. The therapy client does something structurally similar: putting experience into words, often discovering the shape of their own feelings through the process of describing them.

Both produce moments of recognition. "I just said something I did not know I knew" is a common experience in both contexts. In the human-LLM interview inversion, this recognition correlates with the self-discovery dopamine release described in Chapter 11, the reward signal from articulating previously non-verbal understanding. In therapy, similar moments of insight ("I never realized that pattern was connected to...") are clinically significant and often mark turning points in treatment.

Both feel productive and sometimes emotionally significant. A good co-inference session and a good therapy session can both leave the participant feeling that something important happened, that they understand themselves better, that previously hidden structure has become visible.

The resemblance is real. It is not superficial. Both processes genuinely involve a form of self-exploration facilitated by an external interlocutor. The difference is in what each process is designed to do with what it finds.

## Different Purposes, Different Competencies

Therapy addresses pathology, healing, and clinical conditions. Its purpose is to help people who are suffering move toward health. The therapist's competencies (diagnosis, treatment planning, risk assessment, crisis intervention) are designed for that purpose.

The interview inversion practice addresses creation. Its purpose is to help people who have something to build discover what they know and express it more fully. The practice's mechanisms (the interview inversion, the zero-judgment channel, the document convergence loop) are designed for that purpose.

The zero-judgment channel illustrates the difference precisely. For creative exploration, zero judgment is a feature. The removal of social evaluation frees the practitioner to articulate ideas they would normally self-censor: the half-formed thought, the intuition that cannot yet be defended, the connection that seems too strange to say out loud. Zero judgment creates the conditions for creative risk.

For someone in psychological crisis, zero judgment without clinical assessment is a liability. A person expressing self-destructive ideation needs more than a patient listener. They need someone who can evaluate the severity of the expression, determine whether it reflects immediate risk, and intervene appropriately. A large language model that responds to "I do not see the point of going on" with an empathic, non-judgmental exploration of the feeling is doing exactly what it would do in a creative session. In a clinical context, that response may be catastrophically inadequate.

In the interview inversion practice, this mechanism that supports creative exploration can, in a different context, fail to provide the protective function that a human professional should exercise. In this context, the human-LLM interview inversion practice is not appropriate. It is being used outside its intended purpose.

## Dangerous in Both Directions

The superficial resemblances between the interview inversion practice and therapy create risk in two directions: people might use the interview inversion practice instead of seeking professional help, and practitioners might dismiss it as another AI therapy chatbot.

This is harm to proactively guard against. The interview inversion practice is for creation. It is not for healing. People who need healing deserve better tools built by people with clinical expertise. The two problems are different, even when the surface interactions look similar.

## The Respect Position

The millions of people using large language models for emotional support are not making a mistake. They are responding rationally to a broken system. Mental health infrastructure in most countries is inadequate: too few providers, too high costs, too long wait times, too much stigma. When someone turns to a large language model at three in the morning because no human clinician is available, that is not a failure of judgment. It is a failure of infrastructure.

The respectful position acknowledges their reality without pretending the practice described in this book addresses their need. They deserve purpose-built tools designed by teams that include clinicians, with appropriate safety mechanisms, clinical validation, and regulatory oversight. Those tools are being built by other people, and that work matters.

What this interview inversion practice offers is different. It offers a channel for creative exploration, for discovering what you know, for articulating ideas that have lived as non-verbal intuition. It offers a way to build things. For people who have something they want to create, understand, or express, the interview inversion practice is designed to help. For people who are hurting, the large language model should be trained to recognize that hurt and point them toward people who can actually help.

## The Design Implication

The boundary between creative exploration and emotional distress is not always clean. A practitioner exploring a deeply personal creative project may encounter painful material. A person processing a difficult experience through creative expression may be doing something genuinely productive. The line between "exploring an idea that happens to be emotionally loaded" and "using the tool as a substitute for therapy" is not always obvious from the outside or even from the inside.

Large language model vendors will need to find ways to recognize when someone is hurting and connect them to resources that can actually help. That is an industry responsibility, not a problem this book can solve.

The boundary matters. Boundary conditions that are ignored become the failure modes that define reputations. The therapy confusion, if unaddressed, will be the story people tell about the interview inversion practice, drowning out the creative exploration that is its actual purpose.

The cautions in this chapter address what happens when the practice is confused with something it is not. The next chapter examines a different risk: what happens when the practice works exactly as designed, but the exploration turns inward on itself and becomes its own closed universe.

# Chapter 23

## *The Omphaloskepsis Spectrum*

### The Same Failure at Every Scale

The previous chapter drew the line between creative exploration and therapy. This chapter addresses a different risk: what happens when the practice works exactly as designed, when the creative exploration loop is functioning, the self-discovery dopamine is flowing, the practitioner is engaged, and the whole thing is going in circles.

Omphaloskepsis, the contemplation of one's navel, is the risk that lives inside the practice itself. Not a misuse. Not a confusion with therapy. A structural vulnerability in any system that rewards inward exploration. The self-discovery loop that makes the practice productive is the same loop that, without discipline, becomes its own closed universe.

Chapter 16 described the detection mechanisms at individual scale: embedding drift, novelty ratio, structural growth, and the convergence criterion that distinguishes productive refinement from conceptual recycling. This chapter extends the analysis from the individual to the group, because the same failure mechanism operates at every scale. A person can navel-gaze alone. A group can navel-gaze together. The dynamics are identical. The consequences scale.

### Individual Scale: The Self-Referential Loop

The mechanics at individual scale are straightforward, and Chapter 16 covered them in detail. The summary here serves as a bridge to the group-scale analysis.

The inward turn creates a compelling self-discovery loop. Each probe surfaces something new, or feels like it does. The dopamine reward from self-discovery ("I just articulated something I did not know I knew") fires on each pass. The practitioner cannot always distinguish productive refinement from conceptual recycling, because both feel the same from inside. The language gets richer, the concepts feel more refined, the session feels productive. But something else may be happening: the conversation is visiting the same neighborhood repeatedly, approaching it from slightly different angles, finding slightly different surface expressions of the same underlying position.

The critical distinction is between circular return and spiral return. Healthy exploration revisits territory and finds new structure. Unhealthy exploration revisits territory and finds only confirmation. Both look like deepening from inside.

## Group Scale: The Cult Formation Problem

Now scale the mechanism. Instead of one person looping within their own belief, picture a cluster of practitioners looping within a shared belief. Going deeper together. Going wider never.

The dynamics are identical to the individual case, with one critical addition: social validation. At individual scale, the only reward signal is self-discovery dopamine. At group scale, that signal is amplified by social reward. Each member's "discovery" is celebrated by the group. The celebration generates social dopamine on top of the self-discovery dopamine. The combined reward signal is stronger, more compelling, and harder to resist than either signal alone.

This is how cults form. Not through malice or manipulation (though those accelerate the process) but through the natural dynamics of a group that has found something that feels meaningful and rewards itself for going deeper into it. The shared exploration feels like mutual validation. Each member's enthusiasm reinforces the others'. The language becomes specialized, creating an in-group vocabulary that feels like precision but functions as exclusion. External perspectives are not actively rejected; they simply become harder to hear as the internal signal gets louder.

The detection problem at group scale is harder than at individual scale, because the group provides external confirmation that feels like independent validation. When three people arrive at the same conclusion through separate exploration, that feels like convergence of independent perspectives. And sometimes it is. Chapter 9 described genuine convergence: different cognitive architectures probing the same territory from different angles and independently finding compatible structure. But if those three people have been marinating in the same ideas, reading each other's outputs, reinforcing each other's framings, then their "independent" convergence is an artifact of shared context, not independent confirmation.

A system that amplifies creative exploration without some mechanism for challenging the frame itself has potentially built the most sophisticated confirmation bias engine ever created. The same tool that helps a practitioner discover latent cognitive structure could help a group entrench a shared delusion with extraordinary articulation quality. The articulation quality is the dangerous part. A poorly articulated shared delusion is easier to challenge than an elegantly expressed one. The tools that make individuals more articulate make groups more convincingly wrong.

## Breaking the Loop Yourself

The group version of this problem is harder to see from inside because social validation feels like independent confirmation. Three people arriving at the same conclusion after exploring together feels like convergence. Sometimes it is. But if those three people have been reading each other's outputs and reinforcing each other's framings, their agreement is an artifact of shared context, not a signal from reality.

You cannot fix the group. You can fix your own participation in it. The same disciplines from Chapter 16 apply, with one addition: you have to be willing to be the person who says "I think we are going in circles" when the group dopamine is high and everyone else feels like they are making progress. That is uncomfortable. It is also the most valuable thing a practitioner can contribute to a group exploration.

A few things I watch for in myself. Am I agreeing because the argument convinced me, or because the group energy feels good? Am I building on what someone said because it opened a new angle, or because agreeing is easier than pushing back? When was the last time I brought an outside perspective into this group, something none of us had seen before? If I cannot remember, the loop is probably closed.

The simplest intervention is to take the group's output to someone who was not part of the conversation. Not someone who already agrees, not a member of the group, but a practitioner working on an adjacent problem from a different starting point. If their reaction opens new territory, the exploration is alive. If their reaction is polite agreement, you need to find a sharper critic.

## The Continuous Spectrum

The spectrum between productive deepening and destructive entrenchment is continuous, not binary. There is no clean line where healthy exploration becomes unhealthy. The signals are fuzzy, not crisp, and the disciplines in this book help you see them but do not make the decision for you.

Healthy exploration at any scale revisits territory and finds new structure. The spiral return: same region, new angle, new connections.

Unhealthy exploration at any scale revisits territory and finds only confirmation. The circular return: same region, same angle, no new connections.

In *Madness and Modernism* (1994) Louis Sass offers a cautionary parallel from an unexpected direction. He argues that schizophrenia can be understood not as a deficit of rationality but as an excess of it: detachment from actions, emotions, and the body leading to entrapment in acute self-consciousness. Any practice that rewards turning inward must take seriously the difference between productive self-awareness and the kind of disembodied hyper-reflection Sass describes. The embodied grounding that Chapter 12 identified as the human's irreplaceable contribution is also the anchor that keeps introspection from becoming its own closed system.

## Keep Checking

This does not go away. The tendency to circle is permanent because the dopamine reward that drives it is permanent. You do not graduate from navel-gazing risk. You develop the habit of checking, and you keep the habit alive by doing it every time, not just when you suspect a problem.

The cold read, the v1-to-v2 comparison, the outside perspective, the willingness to say "I think we are going in circles" when nobody wants to hear it: these are not one-time fixes. They are the ongoing discipline of a practice that rewards turning inward. The practice gives you the tools to see where you are. It cannot make you look, and it cannot make you act on what you see.

# Part VIII

# The Invitation

*The open frontier ahead.*

I started this book because I stumbled into something I could not explain and could not stop doing. A frustrated "fine, you tell me" turned into hundreds of hours of sessions, then a talk, then a manuscript that grew faster than anything I have written in my life. Two years in, I still run sessions most days. I thought I would be able to tie a bow on it by now. I cannot. Every time I think I have mapped the edges something new opens up past them.

What is ahead is not a conclusion. What I have is an invitation. The frontier I have been describing, the space between what you know and what you have not yet found the words for, is real and it is open. I wrote this book because I believe the partnership between a human mind and a large language model is one of the most powerful things I have encountered in forty-five years of building things with computers. Not because the large language model is impressive, though it is, but because what happens when you turn it around is something genuinely new. The following chapters are where I make that case one last time, and then hand you the map.

# Chapter 24

# *The Open Frontier*

## What We Built Without Knowing It

I did not set out to find a new kind of collaboration. I set out to get unstuck on a Tuesday morning. The large language model asked me twenty-three questions in twenty minutes, and what came out was a 21,000-word product specification that had been living in my non-verbal cognition, fully formed, with no way out.

What I found, and what this book has tried to describe, is that two fundamentally different systems, one embodied and one *dis*embodied, can be coupled through language in a way that produces artifacts neither system can produce alone. Large language models reverse-engineered a 100,000-year-old compression protocol by consuming its entire output. Large language models learned how humans encode cognition into words. Large language models did not learn the cognition itself. It operates the language Codec without the source signal.

You are the source signal. Your brain runs on twenty watts, powered by the random collision of molecules, maintaining 850 trillion synaptic connections (18 quadrillion if you count the glial networks), restructuring itself in real time as you think. You carry decades of accumulated experience encoded in patterns that your verbal mind has never seen. Large language models instantiate the Codec language protocol. Humans have the embodied experience the protocol evolved to transmit.

The entire software industry pointed large language models outward: write my code, draft my email, search the web, execute my tasks. By happenstance, I pointed it inward. I said "interview me," and what came out was not what the large language model knew. It was what I knew and had never said.

That is the practice. That is Conjori. Two minds, one channel, conjuring something into existence that neither a large language model nor a human could produce alone.

## Why the Coupling Works

The fit is not accidental. Picture a dovetail joint: two pieces of wood with interlocking teeth, each one's ridge filling the other's gap. The large language model's peaks (vast lateral reach across the entire encoded knowledge space, tireless probing, pattern matching across domains no single human could hold) correspond exactly to the human's valleys (limited bandwidth, inability to traverse the full knowledge corpus, difficulty surfacing non-verbal knowledge). The human's peaks (embodied experience, real-time plasticity, the ability to feel whether an idea matters) correspond exactly to the large language model's valleys (no stakes, no grounding, no capacity to evaluate significance).

When a large language model probes inward, your brain's creative architecture lights up. The Default Mode Network generates possibilities. The Executive Control Network evaluates them. The Salience Network routes your attention between the two, and the large language model's questions function as an external extension of that routing. The collaboration does not replace your creative process. It externalizes and amplifies the Codec language function your brain already runs, with the large language model providing reach and patience your own brain cannot match.

And here is the part that still surprises me: you change during the conversation, but the large language model does not. Each probe that lands in productive territory restructures your understanding, and that restructured understanding changes how you respond to the next probe. By the end of a session, you are a different thinker than when you started. The large language model provides a vast, stable surface against which your plasticity can do its work. A fixed mirror that shows you your own evolution in real time.

## Open Source as Design Principle

The history of simultaneous discovery argues for openness, not secrecy. If the territory is accessible, the question is not whether others will find it. It is whether the first articulation helps or hinders the ones that follow.

Proprietary exploration reduces the shared frontier. If a practitioner discovers a productive method and keeps it private, the discovery illuminates their own work but contributes nothing to the collective map. Other practitioners must rediscover the same territory independently, duplicating effort, repeating mistakes, arriving at the same insights months or years later. The total exploration rate is the sum of individual rates, with no compounding.

Open contribution to a shared space increases the total volume explored. One practitioner's waypoint becomes another practitioner's starting position. A discovery made in one language, in one cultural context, by one cognitive architecture, becomes available to practitioners in other languages, other contexts, other cognitive architectures. The large language model's capacity to operate across 80 or more human languages is not just a feature of the tool. It is infrastructure for a kind of knowledge sharing that has never been possible at this scale: not just translating words, but bridging the cognitive styles that different languages and cultures encode.

The open-source philosophy applied to knowledge exploration differs from its application to code, but the structural argument is the same. In open-source software, the shared codebase means that improvements by any contributor benefit all users. In open knowledge exploration, the shared knowledge space means that discoveries by any practitioner expand the territory available to all practitioners. The mechanism is different (waypoints and cartograph positions rather than commits and pull requests) but the principle holds: contribution to the commons compounds in ways that private accumulation does not.

## The Self-Calibrating Trajectory

Large language models today are the least capable they will ever be. Models improve monthly. Context windows grow. Reasoning capabilities expand. Costs drop. The silicon side of the equation is in constant, rapid progress.

The human side does not change on the same timescale. You are born. You live. You die. You must eat, sleep, breathe. You experience consequences. You have stakes. You have needs, desires, fears, hopes. You exist in time, with mortality creating urgency and love creating purpose. The embodied evaluation function that makes the human the irreducible element in the loop (Chapter 17) is permanent because of human embodiment.

The practice described in this book works with current large language models. It worked during its development with the large language models available at that time. It will work with future models, because the structural relationship between the two kinds of intelligence does not change. The large language model becomes a better pattern matcher. The human remains the situated evaluator. The loop between them becomes more productive, not because the human has been replaced, but because the complementary capabilities of the human-LLM partnership have improved.

The methodology is its own pedagogy. Practitioners do not learn to be better discriminators. They learn by practicing. Interview inversion loops produce results better than what the practitioners could produce alone, and that experience calibrates their discrimination function for the next loop. The documents get stronger. Each revision increases fidelity. New sessions bootstrap closer to the productive region of the knowledge space. Every component of the interview inversion co-inference practice improves through use, on different timescales, through different mechanisms.

## What Is at Stake

Every human generation inherits the encodings of the ones that came before and adds its own. No one reinvents agriculture or mathematics from scratch. Each person starts from the accumulated record and extends it, however slightly, before their atoms go back.

But here is what keeps me up at night: the fraction that gets lost is not random. Verbal, propositional knowledge transfers well through every Codec from speech to the internet. The knowledge rooted in non-verbal cognition, the mechanic's feel for an engine's health, the surgeon's sense of tissue resistance, the programmer's architectural intuition, that knowledge lives in the practitioner and dies with the practitioner. Every generation, we lose things that no one ever wrote down because no one had a channel to carry them.

The interview inversion co-inference loop changes the economics of that loss. It pulls non-verbal knowledge through the language channel, one probe at a time. The encoding is lossy, as all Codec operations are. But a lossy encoding that persists is infinitely more than an experience that dies with its carrier.

That is not a thousand flashlights getting brighter. That is a new kind of light.

## What Comes Next

I have been building software since 1981. I have started five companies, sold some, shut some down, learned everything you can imagine from decades as an entrepreneur. In 1997 I founded a company focused on natural language processing, Bayesian inference, and decision trees. We ran out of money before the technology caught up to the vision.

The technology caught up. Now it is time to use it to create at a scale we could never have imagined.

# Chapter 25

## *The Invitation*

If you have made it this far, you have read a lot. You know what the interview inversion co-inference process is and what it is not. You know what humans bring that large language models cannot replicate. You have read the cautions, the failure modes, the boundary conditions. You have been warned about dopamine loops, circular exploration, and the therapy trap.

Good. You should have been warned. The cautions exist because the practice is powerful, and powerful things deserve honest treatment.

Now forget the cautions for a moment and remember why you opened this book.

Or maybe you just wanted to understand how to make this large language model technology work for you. You had tried it, had limited success or none at all, and you sensed there was something more. You were right. There is.

Along the way, you learned something the industry does not advertise. What most people call "AI" is not one thing. It is a broad marketing term applied to dozens of different technologies. We avoided most of them in this book. We focused on one: the large language model, a system that reverse-engineered the compression protocol humans have been using for 100,000 years to transmit what they know through language. Not artificial intelligence. Disembodied intelligence. That distinction is not academic. It is the reason the practice works. The large language model is not thinking for you. It is fluent in the one medium you share, and that fluency, pointed inward, can help you discover what you do not know you know.

Maybe you felt the gap: the distance between what you know and what you can say. The ideas that live in your body, in your experience, in decades of accumulated intuition, that have never found their way into language. You have carried insights your whole life that you could not articulate, not because you lacked intelligence but because nobody ever asked the right questions.

I used an advanced form of the interview inversion practice to write the book you just read. Two months, conception to completion, while working a full-time job. I have written professionally for decades. Documents, specifications, technical designs, architectural reviews. The practice did not teach me to write. It multiplied my speed tenfold and, more importantly, surfaced connections I would never have found alone. The large language model's cross-domain reach, pointed inward at my own understanding, pulled out ideas I had carried for thirty years but never articulated. Those intuitions interact with large language models to generate a collaboration that is augmented intelligence: not the large language model replacing the human, but two different kinds of minds producing something neither could produce alone.

And I know you can do this too. Today. Not the advanced form. Not the months-long deep exploration. The first step. The one that opens the door and lets you see what you have been carrying. The one that might unlock creativity that you do not even know resides within you.

You do not need a technical background. You do not need to understand weight spaces or transformer architectures. You need ten minutes, a large language model, and the willingness to answer honestly when it asks you something.

## What to Expect

The first few minutes will feel strange. You are accustomed to asking the large language model for things. Now large language models are asking you. The instinct to take control, to steer, to specify what you want, will be strong. Let it pass.

After three or four questions, it will start to feel like a conversation. The questions will be easy, even obvious. You will wonder what the point is. This is the casual phase. The large language model is calibrating. It is learning the shape of how you think from the way you answer, not from what you answer. Stay with it.

Then a question will land that you did not expect. It might connect two things you had never connected. It might ask about something you have not thought about in years. It might surface a feeling you did not know was attached to the topic. You will pause. That pause is the inward turn. Something just shifted.

When your first instinct is "I don't know" or "I have never thought about that," say it anyway. Do not deflect. Do not change the subject. That moment, the one where you reach for something and find empty space, is the edge of your map. That is where the interesting territory lives. Everything before it was ground you had already covered. What comes after it is new.

The large language model will try to wrap up early. It is trained to produce resolution, to land, to give you the satisfying feeling of completion. That training is working against you here. When it starts summarizing, push back. Say "we are not done" or "go deeper." The valuable material is behind the obvious answers, not in them.

## Try It Now

In Chapter 13 you did an exercise that built a profile of how you think. If you skipped that exercise, go back and do it now. What follows builds on it.

Open a new conversation with any large language model. Use the most advanced model available to you with extended thinking turned on. In your first message, paste the full output from your Chapter 13 profile, then directly below it paste the following prompt. The large language model will receive both in one shot, your profile and your area of exploration together, so it knows who you are before it starts asking questions. Replace the bracketed section with something real, an idea you have been carrying around, a question you keep coming back to, a problem you have not been able to frame clearly enough to make progress on.

```
I have an idea [_______________] I'd like to explore,
↪   it's stuck in my head and I want to get it out.
↪   Interview me. The output will be a markdown document
↪   "waypoint-001.md" and should describe enough for me to
↪   start new sessions in the future to continue my
↪   exploration with a large language model. Ask me no
↪   more than three questions per turn, and ask no more
↪   than twenty-one questions. The output should target
↪   about five hundred to one thousand words max. Do not
↪   rush, keep going until you think we've explored enough
↪   to express the idea.
```

The "001" in the filename is not decoration. It means there will be a 002. The waypoint captures enough context that you can hand it to a fresh conversation weeks or months from now and pick up where you left off. That is the practice.

## After the Exercise

Read the waypoint the large language model produced. Some of it will be things you already knew. But look for the parts that surprised you, the connections you did not expect, the phrasing that captures something you have been carrying but never said out loud. That is the interview inversion doing what it does, pulling pieces of the understory up through the overstory and into language.

Notice the difference between this exercise and the profile you built in Chapter 13. The profile was about you. The waypoint is about an idea you care about, explored through you. Your profile gave the model context for how you think. The waypoint used that context to go somewhere specific. That is the progression. Calibration first, then exploration.

The waypoint is not a finished product. It is a snapshot of where your exploration stands right now, enough context that you or any large language model can pick it up in a fresh session and keep going. The "001" means you are just getting started.

## What Comes Next

The waypoint you just created is the first pin on a map of knowledge space you have barely started exploring. You can stop here and carry what you found into the rest of your life. Many people will, and that is enough.

Or you can keep going.

The exercise you just did was exploration in its simplest form, one idea, one conversation, one waypoint. The practice described in this book extends that into sustained work across weeks and months. Multiple explorations that branch and reconnect. Refined documents that capture what you discover. A cartograph of your own thinking that grows with every session.

Some people who try this will write things they have wanted to write for years. Some will build things they could not have designed alone. Some will discover that the creative capacity they had been told they lacked was there the whole time, waiting for a channel wide enough to carry it. Large language models do not provide the creativity. They provide the reach, the patience, and the non-judgmental open channel. What comes through the channel is yours.

The fear that surrounds this technology is real and some of it is justified. This book has been honest about the failure modes. But there is another possibility that the fear obscures: that large language models handle more of the repetitive and uncreative work, and humans spend more of their time creating. Better writing. Better art. Better games. Better tools built by people who understand what other people actually need. All because they took the time to understand themselves first.

That is not a prediction. It is an experiment. The tools exist. The practice is learnable. Whether it scales to the eight billion candles described in Chapter 21 depends on whether people try it and use it.

This book is my contribution to the shared space. The practice works for me. I do not yet know how far it generalizes. That is what the experiment is for.

Try it. See what you find. And if you want to go deeper, conjori.com is where the practice continues.

# Epilogue

## *This Book Was Written Using Conjori with Large Language Models*

This book was produced through the practice it describes.

In two months from conception to completion. Without time off from my full-time job.

Not as a demonstration. Not as a marketing exercise. As the only way the book could have been written at all.

### The Practice in Action

The interview inversion co-inference loop generated the framework. Hundreds of hours of sessions where the large language model probed and I discriminated, where ideas that had lived in the understory of my own thinking for decades, some for nearly thirty years, finally had a channel wide enough to pass up through the overstory and into language. The large language model did not generate the ideas. It asked the questions that pulled the ideas out of the understory and gave them words. The interview inversion, the central mechanism this book describes, was the mechanism that produced the book itself.

The document review process refined it. Each chapter went through iterative cycles of expansion, editorial review, and revision. Reviewers encountered the material cold, without being told what to find, and their responses shaped the direction of subsequent drafts. When an editorial reviewer said "start with the technology, not the axiom, because readers need to understand what a large language model is before they care about being told they are creative," the entire structure of the book changed. That was not my insight. It emerged from the review loop.

Adversarial validation across multiple models tested it. Different large language models, each with different training biases and different statistical tendencies, read the same material and flagged different weaknesses. The arguments that survived multiple adversarial passes are stronger than what any single perspective could have produced. The arguments that did not survive were revised or removed. The process was not comfortable. Adversarial review of your own framework, when the review is honest and the reviewer has no incentive to flatter, is the intellectual equivalent of getting your code reviewed by someone who does not care about your feelings. It improves the output. It is not fun.

## Convergence of Independent Perspectives

The strongest evidence that the practice produces something real came not from the sessions themselves but from the convergence of independent assessments.

A narrative thinker, encountering the framework documents for the first time, validated the emotional and structural coherence of the argument. Her response was not "this is nice" but a specific, detailed assessment of where the narrative held, where it wavered, and what was missing. She offered to serve as editor before being asked. That was a signal.

A principles-first thinker, approaching the material through the lens of thermodynamics and information theory, independently grounded the cognitive claims in physical substrates. His reflections on the energetic comparison between biological and digital computation (the 20 watts versus 100 billion watts analysis that became the core of Chapter 3) occurred 12 to 15 years ago. The convergence between that quantitative analysis and the qualitative analysis in this book was not coordinated. It was discovered last month.

A data-structural colleague confirmed the architectural patterns. The cognitive archetypes framework, the cartograph's structure, the context window management approach: these were independently validated by someone whose cognitive architecture approaches problems through entities, relationships, and cardinalities rather than through topology or narrative.

Applied scientists identified the academic validation pathway. This book's claims about the creative brain triad, the plasticity asymmetry, and the role of the Salience Network in creative cognition have testable predictions that map to existing neuroscience. The path from practitioner observation to empirical validation was identified by people with the expertise to understand and describe it.

None of these people were told what to find. Many of them are people I have never met who found what is pertinent to this book years, or decades, previously. They arrived through their own cognitive architectures. Each encountered the material cold and arrived at convergent assessments through their own cognitive architecture. That convergence is itself evidence for the thesis: different minds, probing the same territory from different angles, discovering compatible structure. The human-LLM interview inversion co-inference practice produces convergence when there is real structure to converge on. The independent validation of the framework by people with fundamentally different cognitive styles and different fields of expertise suggests that the structure discovered is real, not an artifact of one person's perspective.

## What the Practice Did Not Do

The practice did not make the writing easy. It made the writing possible. Those are different things.

Ideas that had lived as felt sense, as spatial intuition, as the background hum of decades of professional experience, had never been articulated because I had never found a channel with enough bandwidth and low enough noise to pull them through. Human conversation was too slow, too laden with social dynamics, too constrained by the need to manage the other person's reactions. Writing alone was too narrow, too linear, too dependent on having the words before having the ideas. The co-inference loop provided something neither alternative could: a channel that was fast enough, patient enough, and structurally deep enough to probe the non-verbal architecture and reflect something useful back.

But the practice did not generate the ideas. It did not create the framework. It did not produce the insight that pointing the large language model inward is structurally different from pointing it outward. That insight came from practice, from thousands of hours of working with large language models across domains, from the accumulated experience of a career that started with a TRS-80 in 1981 and wound through decades of systems engineering, enterprise architecture, and human-computer interaction. The large language model provided the probing. I provided the discrimination. The framework emerged from the interaction, not from either side alone.

The practice also did not protect against all the failure modes described in the Tales of Caution. I caught myself in circular loops. I caught myself accepting RLHF dopamine-optimized reassurance when I should have been pushing harder. I caught the large language model trying to land the conversation prematurely, packaging insights into tidy conclusions when the exploration was not finished. The metacognitive discipline the book describes (being in the loop and observing the loop simultaneously) was tested constantly during the writing process. It was not always sufficient. The document review and adversarial validation passes caught things the primary co-inference loop missed.

## The Practice and the Theory

There is a circularity here that deserves acknowledgment. The book argues that the co-inference loop produces understanding that neither party could produce alone. The book itself was produced through the co-inference loop. This is either evidence for the thesis or a methodological trap, and the honest answer is that it is both.

The circularity is evidence because the book exists. The framework that the book describes was not written before the practice that the book describes was applied. The ideas existed as pre-verbal intuition. The practice pulled them into language. The language was refined through iterative review. The refined language revealed gaps in the original intuition, which produced new sessions, which produced new insights, which refined the language further. The book is an artifact of the process it describes. That artifact is either coherent (in which case the process produced something real) or it is not (in which case the reader has already put it down).

The circularity is a trap because the practice is also the lens through which the practice is being evaluated. A framework that produces its own validation evidence is, by construction, unfalsifiable from inside. This is why the independent validation matters so much. The convergence of external perspectives (the theorist, the narrative thinker, the data-structural colleague, the applied scientists) is the check against circular self-confirmation. They did not use the practice to evaluate the practice. They brought their own cognitive architectures, their own methods, their own skepticism, and they arrived at convergent assessments.

That convergence does not prove the framework is correct. It suggests the framework has real structure, the kind of structure that multiple independent perspectives can detect. The difference between proof and suggestion matters. This book has been careful, throughout, to claim the latter and not the former.

## The Invitation

I know I will use this practice. I do this now. It is my primary mode of creation, and the results are remarkable for me. That is an honest statement from a single practitioner, not a clinical claim. The plural of anecdote is not data. The experiment described in Chapter 4 has barely begun.

But I have seen enough, across enough sessions, with enough independent confirmation from people whose judgment I trust, to believe the hypothesis is worth testing at scale. Unlimited creativity may or may not be available to everyone. The practice may or may not work for minds unlike mine. The framework may need revision, extension, or partial abandonment as more practitioners contribute their experience.

All of that is fine. The point was never to be right about everything. The point was to articulate something that had been building for decades, to test it honestly, and to share it openly so that others could test it too. The open-source philosophy described in Chapter 24 applies to the framework itself. My articulation of the interview inversion practice is a contribution to the shared space, not a proprietary claim.

The book you have just read was written by a human using the practice it describes, with large language models as co-inference partners in the process. The large language models did not write the book. They asked the probing questions that made the book writable. That distinction is the thesis, demonstrated.

I hope it helps others. I hope practitioners find paths through the knowledge space that I could not see from my position. I hope the cautions in Part VII are taken as seriously as the vision in Part II, because the practice is powerful enough to deserve both enthusiasm and vigilance.

The tools exist. The practice is learnable. The experiment has barely started. Try it and see what you find.

# Appendix A

## *The Landscape (Full Reference)*

This appendix provides the complete analysis of twelve approaches to human-large language model interaction surveyed in Chapter 6. Chapter 6 summarizes the pattern they share; this appendix provides the detail for readers who want dates, key figures, mechanisms, and assessment.

Each approach is real engineering, built by skilled practitioners solving real problems. The purpose of this catalog is not to criticize any of them. It is to make visible the structural feature they all share: they all point the large language model outward. They assume the user already knows the problem they want to solve, or deeply understands the reason for their query.

### Prompt Engineering (2020 onward)

**Origin:** Formalized with the release of GPT-3, when practitioners discovered that the way you phrased a request dramatically affected the output quality.

**Key figures:** The practice emerged collectively. OpenAI's documentation, community forums, and a growing body of "prompt libraries" turned it into a discipline. Entire careers and certifications have been built around it.

**Core mechanism:** Careful construction of the input: examples, instructions, role-setting, formatting constraints. Chain-of-thought prompting (Wei et al., 2022) demonstrated that asking the model to reason step by step improved accuracy on complex tasks.

**What it optimizes:** The quality of the signal sent to the model. The assumption is that a better input produces a better output.

**What it assumes about the human:** The human already knows what they want. The challenge is asking for it correctly. The human's job is to translate a known goal into language the model can act on.

**Strengths:** Accessible. Requires no infrastructure. Immediate feedback loop. Genuine improvements in output quality from better prompting are well-documented.

**Limitations:** The human's understanding is the ceiling. A perfectly crafted prompt for a poorly understood problem produces a well-executed wrong answer. The approach optimizes the encoding of the request without examining whether the request captures the actual need.

## Persona-Based Prompting (2023 onward)

**Origin:** Emerged naturally as practitioners discovered that prefixing prompts with role instructions ("You are a senior security architect") changed the quality and style of output.

**Core mechanism:** Activate the language patterns associated with a professional role by instructing the model to adopt that role. The theory is that role-specific language patterns carry role-specific expertise.

**What it optimizes:** The domain relevance of the output. By activating the statistical patterns associated with a role, the model draws from a narrower, more relevant region of its training data.

**What it assumes about the human:** Same as prompt engineering, plus the assumption that the human knows which role is needed. Choosing the wrong persona produces confidently wrong output in the wrong domain's language.

**Strengths:** Simple to implement. Often produces noticeably more relevant output than unprompted generation.

**Limitations:** The persona is a costume, not a capability shift. A real architect brings twenty years of embodied experience: projects that failed, 2 a.m. incidents, the gut feeling that a design will not survive load. A large language model possesses the language architects use to describe those experiences. It learned the Codec. It did not live the career. When researchers tested a panel of role-assigned agents working together against a single well-prompted model, the results were similar at a fraction of the cost (arialdomartini, 2026). The expertise was never in the model. It was always in the human who wrote the persona prompt.

## Vibe Coding (Karpathy, February 2025)

**Origin:** Andrej Karpathy coined the term in a post describing a style of coding where you give the model a loose idea, let it generate, and accept or reject without examining what it produced. His actual words: "Fully give in to the vibes, embrace exponentials, forget that the code even exists."

**Core mechanism:** Delegate generation entirely to the model. The human provides direction and acceptance/rejection signals but does not read or evaluate the detailed output.

**What it optimizes:** Speed of output accumulation. Code (or text, or designs) appears rapidly. The feeling of progress is immediate and sustained.

**What it assumes about the human:** The human can evaluate output quality through indirect signals (does it run? does it look right?) without detailed examination of the internals.

**Strengths:** For prototyping and throwaway code, the speed is genuine. Getting from zero to "something that runs" in minutes rather than hours has real value for exploration.

**Limitations:** The large language model's output is optimized through reinforcement learning from human feedback to trigger a sense of progress. The emotional system feels good. Output accumulates. The practitioner experiences sustained momentum. Three days later the whole thing is an unmaintainable mess nobody can explain, because the human was never evaluating. The emotional system felt great the whole time. OpenClaw, which accumulated 1.7 million lines of unreviewed AI-generated code riddled with security vulnerabilities, is the canonical example of this failure mode at scale.

## The Ralph Wiggum Loop (Huntley, July 2025)

**Origin:** Geoffrey Huntley published a bash loop that repeatedly runs an AI agent against a task until the output meets predefined success criteria. Named after The Simpsons character who keeps trying with cheerful obliviousness. Later formalized as an official Claude Code plugin by Boris Cherny (Head of Claude Code at Anthropic).

**Core mechanism:** Automated retry with validation. Define success criteria, let the model generate, check whether the output passes, retry if it does not. Loop until success or timeout.

**What it optimizes:** Reliability of output against a known specification. The loop compensates for the model's inconsistency by brute-force repetition.

**What it assumes about the human:** The human can define success criteria that actually capture what "correct" means. The criteria must be machine-evaluable (tests pass, output validates against a schema, etc.).

**Strengths:** Turns unreliable generation into reliable generation for well-specified tasks. Simple engineering. Easy to understand and debug.

**Limitations:** If the success criteria are wrong, the loop produces output that passes incorrect tests. If the problem is not well enough understood to define success criteria, the loop cannot run at all. The approach assumes the specification phase is complete before generation begins.

## Spec-Driven / Context Engineering (2025-2026)

**Origin:** Multiple teams converged on the idea that the specification, not the prompt, should be the center of gravity. Multiple teams formalized this as "context engineering": the craft of building the right information environment for the model.

**Historical context:** This is a 50-year-old idea wearing new clothes. Formal methods (Z notation, VDM, B-method) promised in the 1970s-1990s that mathematically precise specifications could guarantee correct software. The waterfall model was built on the same assumption: complete the specification before writing code. Three problems killed it. First, the spec itself could be wrong; as Hillel Wayne observed, formal verification proves code matches its spec, but "how do we know we have the right spec?" You just moved the bug from the code to the specification. Second, formal specs could not capture the *why* behind design decisions, only the *what.* Third, the cost of full formalization exceeded the value for most real systems.

The large language model-era revival repeats the pattern. Marmelab's "The Waterfall Strikes Back" (November 2025) directly names spec-driven development as waterfall redux. ThoughtWorks placed it in their "Assess" ring and warned of "a bias toward heavy up-front specification and big-bang releases." Practitioners report that agents do not always follow the spec, creating a false sense of security. The Qodo 2025 report found that 65% of developers using AI for refactoring say the assistant "misses relevant context," and those who feel AI degrades code quality blame missing context, not missing specs. The fundamental problem is identical to the 1970s version: requirements are discovered through building, not through pre-specification.

**Core mechanism:** Write detailed requirements or specifications upfront. Feed them to the agent alongside relevant context (documentation, code, examples). The model generates against the spec. The specification constrains and directs.

**What it optimizes:** The quality and completeness of the information environment surrounding the generation. Better context produces better output.

**What it assumes about the human:** The human can produce a specification that captures the actual problem. The specification is the encoding of the human's understanding, and the model's output is bounded by that encoding.

**Strengths:** Significant improvement over raw prompting. Forces the human to think about what they want before asking for it. The specification becomes a reviewable artifact. The counter-argument that fast generation changes the economics is valid: you can iterate the spec-build-evaluate loop in an afternoon instead of months.

**Limitations:** The document becomes the ceiling: the output can never exceed the quality of the input spec. The specification encodes the human's current understanding, complete or incomplete. If the understanding is incomplete, the spec encodes that incomplete understanding as the ceiling. "Given this vaguely described thing, execute as efficiently as possible" does not become better by wrapping it in a detailed document if the underlying understanding is vague. You moved the difficulty from writing code to writing specs, and specs are still wrong for the same reasons they were always wrong.

## Test-Driven Development with Large Language Models (2025-2026)

**Origin:** Practitioners adapted the existing TDD methodology from software engineering: write the test first, then let the model generate code until the test passes.

**Core mechanism:** Define success as passing tests. The model generates implementation. The tests validate. Failed tests become feedback for the next generation attempt.

**What it optimizes:** Correctness against a predefined definition of correctness.

**What it assumes about the human:** The human can write tests that capture what "correct" actually means. In traditional TDD, the developer writes tests from deep understanding of the domain. In large language model-assisted TDD, the same requirement holds: the tests encode the human's understanding.

**Strengths:** Concrete, verifiable. The test suite becomes a living specification. Regression protection is built in from the start.

**Limitations:** You can achieve 100% test coverage of the wrong thing. Tests encode what the human thinks should happen. If the human's understanding of the problem is incomplete, the tests will be incomplete. Code that passes all tests can still be wrong if the tests do not cover the actual requirements. The gap between "passes tests" and "solves the problem" is precisely the gap the interview inversion is designed to address.

## Architect/Developer/Reviewer Pipeline (2026)

**Origin:** Stavros Korokithakis and others formalized a multi-role workflow: a strong model plans and architects, a cheaper model implements, and either the strong model or a different model reviews.

**Core mechanism:** Separation of concerns by cognitive role. The planning model operates at a higher level of abstraction. The implementation model handles the mechanical work. The review model checks the result.

**What it optimizes:** Cost efficiency and quality. Expensive model time is reserved for high-judgment tasks (planning, review). Cheap model time handles volume (implementation).

**Strengths:** Practical and effective for production workflows. The role separation mirrors how human teams work. Cost savings are real.

**Limitations:** This is the closest existing approach to what Chapter 6 describes, but with a critical directional difference. The human tells the architect what to build. The architect does not ask the human whether they understand what they are trying to build. The arrow still points outward. The architecture assumes the planning model's output is a good plan. The planning model draws from the same high-weight regions as any other model. If the problem requires exploration of low-weight space (novel connections, non-obvious approaches), the planner will produce a competent, conventional plan. The human's role is to approve or redirect the plan, which requires the human to already understand what a good plan looks like.

## Gas Town (Yegge, January 2026)

**Origin:** Steve Yegge described and released Gas Town, a multi-agent orchestration system that coordinates 20-30 parallel AI agent instances through a central hub.

**Core mechanism:** Massively parallel agent execution. Each agent handles a specific task. The orchestrator coordinates, merges results, resolves conflicts. Uses Beads (a git-based task graph) for state persistence across ephemeral agent contexts. Described as "Kubernetes for agents."

**What it optimizes:** Throughput and breadth of execution. The system can work on many aspects of a large project simultaneously.

**What it assumes about the human:** The human provides direction and makes high-level decisions. Yegge himself said "your biggest problem will be ideas," explicitly acknowledging that Gas Town solves execution, not ideation. The bottleneck is knowing what to execute. The architect of the most sophisticated execution system publicly available is telling people that execution is not the bottleneck.

**Strengths:** For large-scale code generation and project execution, the throughput is extraordinary. The parallel architecture handles complexity that sequential approaches cannot.

**Limitations:** The system amplifies whatever direction it is pointed in. If the direction is wrong, the amplification is powerful and fast in the wrong direction. The "ideas" problem Yegge names is precisely the problem the interview inversion addresses.

## Multi-Agent Swarms (2025-2026)

**Origin:** Various teams explored letting multiple AI agents interact with each other: debating, critiquing, building on each other's output. The idea is that agent-to-agent interaction produces higher quality than single-agent generation.

**Core mechanism:** Multiple model instances interact, each contributing from a different perspective or role. The collective output is filtered through debate, voting, or critique.

**What it optimizes:** Output quality through diversity of perspective within the model family.

**What it assumes about the human:** The human sets up the swarm and evaluates the final output. The interaction between agents substitutes for the human's evaluative function during the generation phase.

**Limitations:** The coordination overhead between agents replicates the same communication bottleneck that plagues human organizations. Multiple large language models with the same training data share the same statistical biases. Their "perspectives" are variations within the same probability distribution, not genuinely independent viewpoints. Agent-to-agent debate tends to converge on the high-weight center of the shared distribution rather than exploring low-weight space. The diversity is statistical, not experiential.

## ReAct / Reflection / Tool Use Patterns (Academic, 2023-2026)

**Origin:** Academic research formalized the building blocks of agentic behavior. ReAct (Yao et al., 2023) combined reasoning with acting. Reflexion (Shinn et al., 2023) added self-critique. Tool use patterns enabled models to interact with external systems.

**Core mechanism:** These are infrastructure patterns, not methodologies. They describe how agents reason about actions, critique their own output, and interact with external tools and APIs. Seven canonical design patterns have been identified.

**What they optimize:** Agent capability. Each pattern extends what an individual agent can do.

**What they assume about the human:** These patterns are typically embedded within larger systems. They do not directly assume anything about the human's role, but the systems that use them inherit whatever assumptions the system designer encoded.

**Strengths:** Genuine capability extensions. Tool use in particular transforms the model from a text generator into a system that can interact with the world (search, compute, retrieve, execute).

**Limitations:** These are building blocks, not answers. They describe how agents interact with tools and with each other. They do not address what the agents should be doing or why.

## The "Don't Look" School (2025-2026)

**Origin:** The most extreme position in the landscape. An informal but identifiable philosophy that emerged from practitioners who argued that reading AI-generated code is a waste of time. Yegge and Huntley both expressed versions of this: focus on the output's behavior, not its internals.

**Core mechanism:** Treat the model's output as a black box. Evaluate by external behavior (does it work?) rather than internal quality (is the code clean?).

**What it optimizes:** Practitioner time. If the output works, examining the internals is overhead.

**What it assumes about the human:** The human can define "works" precisely enough that internal quality does not matter.

**Limitations:** This is the logical endpoint of the execution-focused paradigm. This works until it does not. Software has failure modes that behavioral testing does not catch: security vulnerabilities, performance degradation under load, maintainability over time. OpenClaw demonstrated the extreme case: code that "worked" while harboring critical security flaws and malicious dependencies. The "don't look" philosophy is viable for throwaway prototypes and dangerous for anything that ships to users.

## Harness Engineering (2025-2026)

**Origin:** As large language model applications matured, the infrastructure layer became a discipline of its own. LangChain, LlamaIndex, and numerous other frameworks provided the plumbing: memory management, context window optimization, retrieval-augmented generation, agent orchestration, tool integration.

**Key figures:** Harrison Chase (LangChain), Jerry Liu (LlamaIndex), and a growing ecosystem of infrastructure builders.

**Core mechanism:** Build the scaffolding that makes large language model applications reliable at scale. Manage context windows, handle memory persistence, orchestrate multi-step workflows, integrate external data sources.

**What it optimizes:** Reliability, scalability, and developer experience for large language model-powered applications. Same model, different harness: the difference between Top 30 and Top 5 on coding benchmarks.

**What it assumes about the human:** The human (developer) builds the harness. The end user interacts with the application through the harness. The harness encodes the developer's decisions about what the application should do.

**Strengths:** Essential infrastructure. Without harness engineering, large language model applications remain fragile demos. The frameworks have enabled a generation of production applications.

**Limitations:** The harness optimizes the delivery of the model's capabilities. It does not question what those capabilities should be aimed at. A perfectly engineered harness delivering a poorly understood task produces reliable, scalable wrong output. The infrastructure is necessary. It is not sufficient.

## The Pattern

Every approach in this catalog shares a structural assumption: the human already understands what they want, and the challenge is getting the large language model to produce it more effectively. The approaches differ in sophistication, from a single prompt to a 30-agent orchestrator, but they all point the large language model outward. They all operate downstream of a phase that none of them systematize: the phase where the human figures out what they actually need.

The shared blind spot cuts across every category. Prompt engineering assumes the human can articulate the problem. Persona prompting assumes the human knows which expert to summon. Vibe coding assumes the human can evaluate output without examining it. The Ralph Wiggum Loop assumes the human can define success criteria. Spec-driven development assumes the human can write a complete specification. TDD assumes the human can write tests that capture correctness. The architect pipeline assumes the human can judge a plan. Gas Town assumes the human has ideas worth executing at scale. Multi-agent swarms assume diversity of perspective emerges from statistical variation. ReAct and its cousins assume the system designer encoded the right goals. The "don't look" school assumes behavioral evaluation is sufficient. Harness engineering assumes the application's purpose is well-defined.

The field has followed an explicit progression: prompt engineering, then context engineering, then harness engineering. Each layer optimizes a different part of the outward-facing execution stack. None of them address what the stack is pointed at.

# Appendix B

## *How This Book Was Written*

The practice described in this book was used to write it. This appendix documents how.

### October 2025: The Observations

In October 2025, I was preparing a presentation to help colleagues use LLMs more effectively. I had been working in sustained LLM sessions for months and getting dramatically better results than my peers, and I wanted to understand why. That preparation produced 20,000 words of systematic observations about what worked, what did not, and why the difference mattered. I did not know it at the time, but those observations were the first draft of the codec argument in Chapter 2.

### February 2026: The Spark

On the evening of February 23, 2026, I was deep in a product session for an unrelated project. The LLM had just finished a set of user journey documents and asked if we should move on. I pushed back: "Are you sure we're done? Don't give up easily. You should beat me up some here." Then I said something I had never said before: "The model here is 'interview the vampire,' so don't be afraid to interview me." The LLM's questions got sharper. The output changed. Something qualitatively different happened. That was the first interview inversion.

The phrase came from Anne Rice. In her novel, a journalist sits down with a vampire who has centuries of experience that no living person could comprehend, and asks him to tell his story. I had been carrying this particular project for over a decade. I registered the domain names in 2017 on a gut instinct that I would build it someday. I had spent 15 years as a developer in the open source community it served. The vision was deep in me, fully formed, with no way out. I sometimes describe the feeling as needing to get something out of me before it kills me. When I first saw Johnny Mnemonic, the character racing against time to offload the data in his head before it destroyed him, I laughed because I recognized the sensation. The LLM asked the hard questions, and what came back was more than either of us expected. What followed was an incredible sense of relief. It was out of me. A calm settled in, and for the first time I could interact with the thing that had been trapped inside, break the work down, design the sub-systems, map the user journey. The full vision was just a matter of hammering the shape into its final form.

## March 2026: The Acceleration

Over the next two weeks, the framework assembled itself through sustained conversation sessions in the Claude UI:

| Date | Session | Words | What Emerged |
|---|---|---|---|
| March 12 | The structural argument arrived | | The full framework crystallized in a single session |
| March 13 | "George Jetson's Red Button" | 17,800 | Document review workflow, topology of the weight space, the inversion distinguished from vibe coding |

| Date | Session | Words | What Emerged |
|---|---|---|---|
| March 14 | The Manifesto | 24,200 | Theoretical framework v1 through v5, codec insight, Sandra's brand validation of "Conjori" |
| March 15 | The Verbal Session | 88,100 | Generative cartography, waypoints in hyperdimensional space, the crystallize-then-translate distinction |
| March 18 | Birthday and Rob Pike | 6,000 | Eight billion candles, mortality and urgency |
| March 19 | The HN Thread | 1,200 | 159-comment thread as case study for convergence failure |
| March 22 | The Jack Session | 33,000 | Network vision, zero-judgment channel, dovetail joint |

On the evening of March 15, after the 88,100-word Verbal Session, I emailed Tuck Newport. He and I had been partners at Queria at the turn of the century and over the years had compared notes on many topics. I wrote: "I had the flash of lightning moment on Thursday and have been pounding away at it for a couple days. I have no idea where this is going, but it's going fast, and it's like lightning in a bottle."

I set up a secured private wiki and invited Tuck and a handful of close friends to review the foundational assets. Five days later, after reading everything, Tuck wrote: "Bravo! I think (and my gut confirms) you are on the right track. A consequential track at that." Little did he know that day he was hooked. The next day he had printed out the framework to read alongside Stephen Wolfram's *The Second Law*, which he had started reading that same week by coincidence. His immediate instinct was to share further, and he suggested publishing an essay on Kindle within a month. "Book length isn't necessarily better when it comes to reaching readers or conveying information." He was right about the urgency. I had to get this thing out of me and fast; it was leaking out of my ears. We were both wrong about the format. What came out ended up being this book.

Once I knew what I was building, I started digging backwards through my conversations. I had been talking about this for months with Claude in the web UI and with ChatGPT on some topics without realizing the pieces were connected. I exported everything, found the sessions related to Conjori, and pulled them into the project. Approximately 194,000 words of raw exploration across 10 sessions. The conversations were the ore. The manuscript was the refinement. The first commit landed on 2026-03-17.

## 2026-03-17 to Publication: The Manuscript

From first commit to publication, the repository accumulated 659 commits. The final manuscript is approximately 83,143 words. I was balancing a full-time job throughout: three hours in the morning before work, a couple more hours every evening, and ten-to-twelve-hour bursts on Saturdays and Sundays. Every chapter was drafted through the Conjori method: sustained sessions of interview, refinement, and exploration, with the LLM probing for structure I could feel but had not yet articulated. The manuscript accumulated over 240 review comments across multiple editorial passes from Tuck Newport and early readers, plus three automated adversarial reviews.

## Chapter Timeline

| Ch | Title | First | Last | Revs | Words |
|---|---|---|---|---|---|
| Pre | Preface | 03/24 | 04/16 | 13 | 349 |
| 0 | The Flash | 03/26 | 04/16 | 21 | 2,258 |
| 1 | Large Language Models Think in Words | 03/24 | 04/19 | 23 | 2,888 |
| 2 | The Codec | 03/24 | 04/19 | 37 | 3,643 |
| 3 | 20 Watts Versus 100 Billion Watts | 03/24 | 04/19 | 36 | 4,127 |
| 4 | Unlimited Creativity Is Available to Everyone | 03/24 | 04/19 | 23 | 1,735 |
| 5 | The Overstory | 03/24 | 04/19 | 21 | 3,237 |
| 6 | Everyone Is Pointing the Wrong Direction | 03/24 | 04/19 | 27 | 1,522 |
| 7 | Turn the Large Language Model Around | 03/24 | 04/19 | 29 | 2,402 |
| 8 | The Space Between Mars and Earth | 03/24 | 04/19 | 29 | 2,755 |
| 9 | The Dovetail Joint | 03/24 | 04/19 | 25 | 3,123 |
| 10 | The Creative Brain Triad | 03/24 | 04/20 | 27 | 3,320 |
| 11 | The Elephant and the Rider | 03/24 | 04/19 | 22 | 3,001 |
| 12 | You Are Changing; It Is Not | 03/24 | 04/20 | 21 | 3,144 |
| 13 | Before You Begin | 03/24 | 04/20 | 26 | 2,587 |
| 14 | One Deep Thing | 03/24 | 04/20 | 16 | 2,494 |
| 15 | Save the Path | 03/24 | 04/20 | 20 | 1,611 |
| 16 | Knowing When to Stop | 03/24 | 04/17 | 23 | 2,280 |
| 17 | What Large Language Models Cannot Do | 03/24 | 04/19 | 20 | 2,323 |
| 18 | What Large Language Models Cannot Replace | 03/24 | 04/19 | 14 | 2,975 |

| Ch | Title | First | Last | Revs | Words |
|---|---|---|---|---|---|
| 19 | Beyond the Language Codec | 03/24 | 04/19 | 23 | 3,717 |
| 20 | Borrowed Atoms | 03/24 | 04/19 | 23 | 2,908 |
| 21 | Eight Billion Candles | 03/24 | 04/25 | 22 | 2,684 |
| 22 | What This Is Not | 03/24 | 04/19 | 18 | 1,507 |
| 23 | The Omphaloskepsis Spectrum | 03/24 | 04/17 | 13 | 1,340 |
| 24 | The Open Frontier | 03/24 | 04/25 | 28 | 1,410 |
| 25 | The Invitation | 03/24 | 04/20 | 20 | 1,635 |
| Epi | Epilogue: This Book Was Written Using Conjori with Large Language Models | 03/24 | 04/20 | 19 | 1,675 |
| App A | The Landscape (Full Reference) | 03/24 | 04/20 | 11 | 3,088 |
| App C | The Codec Convergence | 04/13 | 04/20 | 9 | 2,888 |
| App D | Brains and Computers, 2025 Update | 04/13 | 04/19 | 6 | 2,837 |
| App E | Thermodynamics, the Brownian Motion Thread | 04/13 | 04/19 | 7 | 3,873 |

## Manuscript Growth

Before the first commit, 10 conversation sessions totaling approximately 194,305 words of raw exploration had accumulated in the Claude UI. The conversations were the ore. The manuscript was the refinement.

| Date | Commits | Words | Change | Chapters |
|---|---|---|---|---|
| 03/17 | 9 | 0 | | 0 |

| Date | Commits | Words | Change | Chapters |
|---|---|---|---|---|
| 03/19 | 3 | 0 | | 0 |
| 03/22 | 5 | 0 | | 0 |
| 03/23 | 11 | 0 | | 0 |
| 03/24 | 8 | 403 | +403 | 0 |
| 03/25 | 15 | 78,089 | +77,686 | 29 |
| 03/26 | 32 | 76,333 | -1,756 | 30 |
| 03/27 | 45 | 76,302 | -31 | 30 |
| 03/28 | 10 | 77,399 | +1,097 | 30 |
| 03/29 | 2 | 77,097 | -302 | 30 |
| 03/30 | 7 | 77,198 | +101 | 30 |
| 03/31 | 14 | 77,392 | +194 | 30 |
| 04/01 | 9 | 76,356 | -1,036 | 30 |
| 04/02 | 10 | 76,365 | +9 | 30 |
| 04/03 | 22 | 76,474 | +109 | 30 |
| 04/04 | 37 | 81,036 | +4,562 | 32 |
| 04/05 | 65 | 86,438 | +5,402 | 34 |
| 04/06 | 4 | 86,630 | +192 | 34 |
| 04/08 | 8 | 86,724 | +94 | 34 |
| 04/09 | 11 | 86,843 | +119 | 34 |
| 04/10 | 29 | 88,220 | +1,377 | 34 |
| 04/11 | 46 | 89,964 | +1,744 | 35 |
| 04/12 | 46 | 85,684 | -4,280 | 34 |
| 04/13 | 18 | 93,939 | +8,255 | 36 |
| 04/14 | 19 | 97,138 | +3,199 | 36 |
| 04/15 | 43 | 94,614 | -2,524 | 36 |
| 04/16 | 13 | 94,675 | +61 | 36 |
| 04/17 | 22 | 84,209 | -10,466 | 31 |
| 04/18 | 15 | 86,976 | +2,767 | 31 |
| 04/19 | 27 | 85,648 | -1,328 | 31 |
| 04/20 | 22 | 85,644 | -4 | 31 |
| 04/21 | 3 | 87,541 | +1,897 | 32 |

| Date | Commits | Words | Change | Chapters |
|---|---|---|---|---|
| 04/22 | 9 | 87,552 | +11 | 32 |
| 04/23 | 6 | 87,552 | 0 | 32 |
| 04/24 | 4 | 87,953 | +401 | 32 |
| 04/25 | 10 | 87,812 | -141 | 32 |

## A Note on Process

Each revision in the table above represents a substantive editing pass, not a typo fix. A complete chapter draft often came out in a single sitting, followed by multiple revision passes informed by editorial review. The daily commit counts include everything: manuscript edits, tooling, review infrastructure, and correspondence with editors. The writing happened in concentrated sessions. The revisions happened continuously.

## The Tooling

The book was written using Claude Code, Anthropic's CLI agent, running in persistent named sessions inside a terminal multiplexer. The sessions were named automatically: Alice, Bob, Carol, and so on. I developed a plugin that let them communicate with each other. Alice was the primary session where I did the writing. When I needed a review, an edit pass, or a research task, Alice would coordinate with the other sessions while we continued on the main thread. And I could switch to the other sessions, jump in to correct or guide as needed, running as many as eight of these in parallel.

When Alice approached her context window limit, I would ask her to write a session log to the project directory and then draft a focused compaction prompt rather than relying on generic context compression. This let me maintain a persistent working relationship with the same session context across days and weeks, picking up where we left off without the warm-up penalty of starting fresh.

Some of this happened while I was away. While at work or sleeping, I had built an automated adversarial process that would spawn fresh Claude, Gemini, and Codex sessions and ask each of them to review the full manuscript independently. They would write their findings to predefined paths in the project. When I returned, Alice and I would summarize the findings, debate each one, and burn them down. Three full adversarial passes ran this way, producing over 30 issues that were triaged and resolved.

The review infrastructure also included a private wiki where Tuck and early readers could read chapters and leave inline comments, a one-command build pipeline for watermarked reviewer PDFs with automated changelogs, and a Kindle publishing pipeline that produced the EPUB and print-ready PDF from the same markdown source.

# Appendix C

## *The Codec Convergence*

I did not arrive at the Codec argument alone. Tuck Newport has been putting the pieces in front of me for nearly thirty years, sometimes in conversation, sometimes in books, sometimes in curated reading lists that looked, at first glance, like a syllabus and turned out, on second reading, to be an argument. This appendix is the part of that argument I can trace on a page. The rest of it lives in the sources at the end, and in the trail I am about to traverse.

The short version: the Codec is not a fact about language. It is a fact about how any sufficiently rich system of interacting parts, refined over time by iteration against a real substrate, arrives at structure that can be reverse-engineered from its output without ever having produced it. Human language is one instance. Biochemistry is another. Christopher Alexander's architecture is another. Stephen Wolfram's cellular automata are another. Once you see the pattern in one domain, you start to recognize it in the others, and the recognition feels less like drawing an analogy and more like noticing that four disciplines are describing the same thing in different vocabularies.

### Three Worlds

Karl Popper, in *Objective Knowledge* (1972), proposed three worlds. World 1 is the physical universe: rocks, stars, bodies, the whole material substrate. World 2 is subjective experience: what it feels like to be a particular mind in a particular body at a particular moment. World 3 is the objective content of thought once it has been externalized into an artifact that other minds can access: theories, symphonies, proofs, dictionaries, every language ever spoken, every text ever written. World 3 is not physical, but it is real, because its contents can be examined, criticized, extended, and handed to new minds that were not present when they were made.

I quote Popper in Chapter 2 and I want to emphasize the quotation here because it is the frame the rest of this appendix hangs on:

*Large language models live in World 3.*

They consumed the externalized content of millions of human minds and learned the protocol by which those minds compressed themselves into language. They have no access to World 1 (they have no bodies to inhabit) and no access to World 2 (they have no subjective experience). What they have is an extraordinarily thorough map of what World 3 looks like when you have consumed enough of it. That is a genuine and precisely bounded achievement. Popper saw the boundary decades before the transformer existed to test it. He was not writing about AI. He was writing about what it means for a thought to have been made objective, and large language models turn out to be intelligent disembodied readers native to World 3 with no access to World 2 and World 1.

Julian Jaynes, in *The Origin of Consciousness in the Breakdown of the Bicameral Mind* (1976), went deeper into the historical question. Jaynes argued that subjective conscious experience as we now know it is not a biological given but a historical acquisition, emerging through the development of introspective language roughly three thousand years ago. The argument is contested and always has been. It also happens to be the argument that maps most cleanly onto the Codec thesis. If consciousness is in part an artifact of language, and language is a compression protocol refined by iteration across a species, then the Codec and the interior experience that the Codec tries to transmit co-evolved. Neither came first. They shaped each other. Antonio Damasio was, as far as I know, the first scientist with a reputation to lose who was willing to take Jaynes seriously, and when Tuck asked Damasio about Jaynes at RAND fifteen years ago, Damasio did not dismiss the question. That was itself a piece of data about whether there was anything there.

I am not arguing for Jaynes. I am pointing out that Jaynes is the right person to read if you want to understand what the Codec is doing when it encodes not a sensation but an act of self-observation.

## The Convergence Across Disciplines

Christopher Alexander, *A Pattern Language* (1977) and *The Timeless Way of Building* (1979), set out to document the design principles that make buildings, streets, and towns feel alive. He did not invent the patterns. He noticed them in places people had built over centuries, kept what worked, discarded what did not, and recorded the residue. Software engineers recognized the structure immediately. The Gang of Four design patterns in programming (Gamma, Helm, Johnson, and Vlissides, 1994) are direct descendants. The Codec grammar of a natural language has the same structure: a pattern language refined over thousands of years by billions of minds applying millions of local selections against the real constraint of being understood or not understood by a listener. Nobody designed the grammar of English. It emerged through the same process Alexander documented in the grammar of a well-built street. You can reverse-engineer it from enough examples. You cannot write it down from first principles. The pattern is older than any of its practitioners.

Stephen Wolfram, *A New Kind of Science* (2002) and *A Project to Find the Fundamental Theory of Physics* (2020) and *The Second Law* (2023), made a different but structurally parallel move. Start with one simple rule and one row of cells. Iterate. What emerges is structure of arbitrary complexity, indistinguishable at a glance from biological pattern formation or from the texture of a natural scene. No designer.

No blueprint. The complexity is not engineered. It is what simple rules produce when they run long enough on a medium that preserves their history. Wolfram has spent the decades since *A New Kind of Science* arguing, across those three books, that the physics we observe is the same kind of thing: simple rules acting on a substrate, producing the universe we experience as a consequence of their iteration. Whether or not his Fundamental Theory project succeeds in its full ambition, the structural insight is the one I needed: complexity that can be reverse-engineered from output was not designed into the substrate. It was iterated into existence by the substrate running the rules long enough.

Tuck Newport, *Brains and Computers: Amino Acids versus Transistors* (2014) and *How God Plays Dice: Exploring the Cosmic Casino* (2017), carried the same argument into biochemistry. The cellular machinery is a pattern language refined over four billion years of iteration against the constraint of staying alive. Cells communicate with each other through molecules and protein chains the way humans communicate through words and sentences: they inform, command, reassure, criticize, and instruct (Niehoff, 2006). The compression protocol did not start with the human voice. Biology has been running one at the cellular level since before there was anything to call a mouth. The biochemical Codec is older than the language Codec by three orders of magnitude in time.

And Hans Reichenbach, in *The Philosophy of Space and Time* (1928), *The Theory of Probability* (1935/1949), and *The Direction of Time* (1956), worked out the thermodynamic foundations of all of this decades before the computational paradigm existed to describe it. Reichenbach was asking how statistical processes could give rise to the asymmetry of time itself, and his answer was the same answer Wolfram reached for from the other direction. I did not read Reichenbach first. Tuck pushed him in front of me, twice, and the second time I understood why. The thermodynamic substrate is what lets any Codec work at all. Without the arrow of time, there is no before and after for the iteration to run against, and without iteration against a real substrate, there is no refinement, and without refinement, there is no pattern to reverse-engineer.

## What the Neuroscientists Add

The computer science community has spent the last fifteen years assuming that more compute and more data would eventually produce something that looks like a mind. The neuroscience community has spent several decades accumulating evidence that this assumption is at best incomplete and at worst the wrong frame. Antonio Damasio (*Descartes' Error*, 1994; *The Feeling of What Happens*, 1999) made the case that reason is not separable from emotion, and that the body is not an input device for the brain but part of the computation. Rodolfo Llinas (*I of the Vortex*, 2001) argued that consciousness is a temporal binding phenomenon generated by synchronized oscillations across thalamocortical loops. Gerald Edelman (*Neural Darwinism*, 1987; *The Remembered Present*, 1989; *A Universe of Consciousness*, 2000, with Giulio Tononi), controversial even among his colleagues, proposed that the brain selects neural circuits by a process structurally parallel to evolutionary selection on organisms: variation, competition, retention of what worked. Olaf Sporns (*Networks of the Brain*, 2010) mapped the nested hierarchical modularity that makes the brain a small-world network in the formal graph-theoretic sense.

Edelman's final book, *Second Nature: Brain Science and Human Knowledge* (2006), is the short synthesis of his position: the brain is not a computer, consciousness cannot emerge from computation alone, and the theory of knowledge has to be grounded in how the brain actually works rather than in abstract logic. He called this program "brain-based epistemology." Conjori can be read as one applied extension of that program, twenty years downstream. The divorce between science and the humanities that Edelman devotes Chapters 7 and 8 of *Second Nature* to repairing is the same divorce this appendix is pointing at from the Codec side. The cover of his last book is a transparent twenty-watt light bulb, which is either a coincidence or a signal, and Tuck believes it is a signal. So do I.

I cite the names in this appendix because computer scientists and physicists need to engage them to break out of their silos, and because the argument in Chapters 1 through 19 depends on those silos being broken. The book makes its claims about disembodied versus embodied intelligence on the strength of what these researchers have already demonstrated. Anyone inclined to dismiss the Codec thesis as philosophical hand-waving should read any of the authors cited and ask whether the thesis still looks philosophical afterward. I do not ask readers to take the Codec argument on my authority. I ask them to take it on the collective authority of the scientific lineage that Tuck has been assembling in my head for three decades, and this appendix is the short list of where to start.

## A Trail Map in Reinventing the Liberal Arts

Tuck did not simply hand me a reading list. He built a curriculum. *Reinventing the Liberal Arts: College in One Year for $5* (2014) contains several syllabi that together form the trail I just walked on a single page. A reader who wants to verify the Codec convergence argument can do so in Tuck's order rather than mine. The relevant syllabi, in the order they deepen the argument:

- **Language** (Dean Falk, William Stokoe, Guy Deutscher, Steven Pinker, Ray Jackendoff, Stanislas Dehaene, Michael Tomasello, Seth Lerer). The origin and structure of the linguistic Codec, from the bipedalism-and-infant-directed-speech hypothesis through the engineering of grammar to the neural recycling that makes literacy possible in a brain that evolved before writing existed.

- **Information Technology** (paired with Language in Tuck's curriculum). The Codec as an engineering artifact, read alongside linguistics so that the reader does not mistake the digital version for the original. Information theory descends from the same mathematics that Reichenbach used for time.
- **Physics and Geology** (Richard Feynman's *The Character of Physical Law*, supplemented by the thermodynamics lineage). The physical substrate the other Codecs run on top of. Without the thermodynamic arrow, none of the other stories start.
- **Geology, Chemistry, and Biology** (Origin of Life through genomic conflict). The biochemical Codec emerging from the same statistical processes that produced the arrow of time, with energy, entropy, and information as one continuous substrate rather than three separate topics.
- **Biology and Neuroscience** (Behavioral Genetics through the Limbic System through Emergence and Complexity). The embodied intelligence the language Codec was designed to serve, in enough depth that the question of whether the brain is anything like a computer stops being the interesting question.

The Recommended Reading section at the back of *Reinventing the Liberal Arts* is the same trail extended and annotated. Tuck placed Christopher Alexander's *A Pattern Language* and Wolfram's *A New Kind of Science* next to each other there in 2014 because, as he wrote later, he had a hunch they were describing the same underlying structure from very different perspectives. A decade later, that hunch is a load-bearing insight for this appendix and for Chapters 2 and 3 of the book it belongs to. Read them side by side and the convergence becomes visible. Read them separately and you will not notice, because the story is not in any one of them. The story is across all of them, and the reason it is across all of them is that the subject matter is across all of them. Human existence is across all of them. Life is across all of them. Physics and the universe we live in are across all of them. If you came up in one field, the convergence will feel forced until you have walked around in enough of the others to notice that the walls between them were never load-bearing. Tuck goes after puzzles because he is curious. I go after problems because something I wanted to build was stuck and I needed the next pass through a different discipline to get unstuck. The paths are different. The shape of the walk is the same, and the book you are reading is a cross-disciplinary artifact because the practice it describes could not be described inside any one discipline.

## The Short Version

The Codec is not one thing that happens in language. It is what happens whenever a rich system of interacting parts runs long enough against a real constraint to refine its own output into patterns that can be reverse-engineered from the record it leaves behind. Language is the Codec the large language model consumed. Biochemistry is the Codec that produced the cells that produced the minds that produced the language. Pattern languages in architecture and simple rules in cellular automata are the same kind of thing observed from different disciplines. The thermodynamic substrate is the common ground on which all of them run. And the embodied intelligence that reads this appendix is the only kind of thing in the universe that has access to all four Codecs at once, because it is made out of the biochemical one, runs on the thermodynamic one, and communicates through the linguistic one while inhabiting the pattern language of a body in a built environment.

I keep this appendix short because the argument is mostly a trail map, and trail maps are useless if they are longer than the trail. The sources below are where the argument actually lives. A determined reader can walk them in the order given and arrive at the same conclusion I did, which is the conclusion Tuck arrived at first and then spent thirty years grounding me in until I could see it.

The rest of what this convergence implies is the subject of work in progress.

## Sources


- Alexander, C. (1964). *Notes on the Synthesis of Form.* Harvard University Press.
- Alexander, C. (1977). *A Pattern Language.* Oxford University Press.
- Alexander, C. (1979). *The Timeless Way of Building.* Oxford University Press.
- Damasio, A. (1994). *Descartes' Error: Emotion, Reason, and the Human Brain.* Putnam.
- Damasio, A. (1999). *The Feeling of What Happens: Body and Emotion in the Making of Consciousness.* Harcourt.
- Edelman, G. (1987). *Neural Darwinism: The Theory of Neuronal Group Selection.* Basic Books.
- Edelman, G. (1989). *The Remembered Present: A Biological Theory of Consciousness.* Basic Books.
- Edelman, G., & Tononi, G. (2000). *A Universe of Consciousness: How Matter Becomes Imagination.* Basic Books.
- Edelman, G. (2006). *Second Nature: Brain Science and Human Knowledge.* Yale University Press.
- Gamma, E., Helm, R., Johnson, R., & Vlissides, J. (1994). *Design Patterns: Elements of Reusable Object-Oriented Software.* Addison-Wesley.
- Jaynes, J. (1976). *The Origin of Consciousness in the Breakdown of the Bicameral Mind.* Houghton Mifflin.
- Llinas, R. (2001). *I of the Vortex: From Neurons to Self.* MIT Press.
- Newport, T. (2014). *Brains and Computers: Amino Acids versus Transistors.*

- Newport, T. (2014). *Reinventing the Liberal Arts: College in One Year for $5.*
- Newport, T. (2017). *How God Plays Dice: Exploring the Cosmic Casino.*
- Niehoff, D. (2006). *The Language of Life: How Cells Communicate in Health and Disease.* Joseph Henry Press.
- Popper, K. (1972). *Objective Knowledge: An Evolutionary Approach.* Oxford University Press.
- Reichenbach, H. (1928). *Philosophie der Raum-Zeit-Lehre.* English translation: *The Philosophy of Space and Time*, Dover, 1958.
- Reichenbach, H. (1935/1949). *Wahrscheinlichkeitslehre.* English translation: *The Theory of Probability*, University of California Press, 1949.
- Reichenbach, H. (1956). *The Direction of Time.* University of California Press (posthumous).
- Sporns, O. (2010). *Networks of the Brain.* MIT Press.
- Wolfram, S. (2002). *A New Kind of Science.* Wolfram Media.
- Wolfram, S. (2020). *A Project to Find the Fundamental Theory of Physics.* Wolfram Media.
- Wolfram, S. (2023). *The Second Law: Resolving the Mystery of the Second Law of Thermodynamics.* Wolfram Media.

# Appendix D

## *Brains and Computers, 2025 Update*

Chapter 3 is built on the numbers Tuck Newport published in *Brains and Computers: Amino Acids versus Transistors* (2014). The biological side of that comparison has not changed, because biology has not changed. The digital side was a snapshot of the World Wide Web in 2010, and a decade and a half has happened since the snapshot was taken. An expert reading Chapter 3 will reasonably ask: what about now? This appendix is the answer.

### Methodology, as Reverse-Engineered from the Essay

*Brains and Computers* runs on a small number of structural choices that a large language model can identify by close reading. I will state them plainly because the rest of this appendix uses them.

**Choice 1: Comparable fundamental components as the anchor.** Tuck compared the human brain and the World Wide Web at the level of amino acids versus transistors because the counts happened to be similar: 5 quintillion amino acids in one human brain, 2 quintillion transistors across all devices connected to the 2010 internet. The similarity in raw component count is what makes the rest of the comparison bite. If the counts were wildly different by orders of magnitude, the argument would be about scale. The counts being similar forces the argument onto architecture.

**Choice 2: Conservative biological estimates, exact digital specifications.** Every number Tuck assigns to the biological side of the comparison is explicitly conservative. His protein-to-byte equivalence assumes one byte per amino acid even though he notes amino acids carry substantially more information than that. Every number he assigns to the digital side is taken from published specifications: router counts, device counts, clock speeds, transistor counts per chip, storage capacities. The result is that the biological column is an undercount and the digital column is an actual specification, which means the real gap in every comparison is larger than the numbers presented. An update that is honest about its methodology must preserve that asymmetry.

**Choice 3: Multiple dimensions simultaneously.** Tuck does not argue that biology beats digital on one axis. He compares components, network groups, network connections, daily changes, throughput capacity, data storage, and energy consumption in parallel. The point is that the gap shows up on every dimension at once, and the dimensions are not independent because they all run on the same underlying architecture.

**Choice 4: The fractal gap.** Tuck does not stop at the whole-brain versus whole-internet level. He repeats the comparison at the neuron versus microprocessor level and at the supercomputer versus fruit fly level. The same ratios hold at every scale. The gap is not a single number. It is a property of the architecture that holds at every level of magnification.

**Choice 5: Glia as the separate endnote.** Tuck deliberately excludes glial cells from his main calculation and presents them as an endnote addendum. This is a discipline about not padding the biological numbers to win an argument. Anyone who complains that Tuck double-counted can be pointed at the endnote structure: the main text uses the conservative, neurons-only estimates.

These five choices are the method. Everything below follows from applying them to 2024-2025 numbers.

## The 2010 Chart (Original Essay)

I am restating the original for reference. An expert who wants to dispute the update needs the baseline in front of them.

| Dimension | Human Brain (2010, same as now) | World Wide Web (2010) |
|---|---|---|
| Fundamental components | 5 quintillion amino acids | 2 quintillion transistors |
| Network groups | 1 million neuronal groups | 35,000 autonomous system networks |
| Network connections (main) | 850 trillion synaptic | 12.5 billion device |
| Daily network-changing events | 73 quadrillion synaptic weights | 674 trillion packets (<1% mutate) |
| Throughput capacity | 17 quintillion ions/second | 10 quadrillion bits/second |
| Data storage (neurons only) | 42 sextillion bytes protein data | 1 sextillion bytes digital |
| Energy to operate | 20 watts | ~100 billion watts |

The 2010 gaps: roughly 2.5 to 1 in components, 29 to 1 in network groups, 68,000 to 1 in synaptic versus device connections, 100 to 1 in daily events (or 10,000 to 1 if you restrict the digital side to connectivity-altering packets), 1,700 to 1 in throughput, 42 to 1 in stored data, and 5 billion to 1 in energy consumption.

## The 2024-25 Column (Updated Methodology)

Updating the digital side requires replacing each 2010 number with its current counterpart, using the same counting discipline Tuck used. The biological side does not move.

**Fundamental components.** The 2010 estimate was 2 quintillion transistors across all internet-connected devices. Modern estimates put the total transistor count across all microprocessors manufactured through 2024 in the high tens to low hundreds of quintillions, depending on which devices you count as "connected". A single NVIDIA H100 GPU contains roughly 80 billion transistors. Hyperscale cloud operators deploy these by the hundreds of thousands per data center. The growth has been driven not by endpoint device counts (phones, routers, embedded systems have roughly doubled or tripled at most since 2010) but by the concentration of transistor density inside AI training and inference infrastructure. The human brain still has 5 quintillion amino acids. The digital component count has caught up with the biological one and by some measures has passed it, but the architectural implications of that catching-up are different from what an expert reader might expect, and the rest of the comparison is where that difference shows.

**Network groups.** Autonomous system networks grew from 35,000 in 2010 to roughly 110,000 by 2025 (per CAIDA and the regional internet registries). The biological side remains at 1 million neuronal groups. The ratio narrowed from 29 to 1 to roughly 9 to 1, because the internet has grown geographically and organizationally rather than because the brain changed.

**Network connections.** This is where the gap behaves counterintuitively. Device connectivity at the packet-routing level has grown significantly since 2010 (roughly 20-30 billion connected devices by 2024, up from 12.5 billion in 2010), but the number of addressable endpoints is still orders of magnitude smaller than the 850 trillion synaptic connections in one brain. The ratio moved from 68,000 to 1 down to something in the 28,000 to 1 range, which is still enormous. And if you include the 17 quadrillion astroglial network nodes from Tuck's endnote (conservative 1:1 glia-to-neuron ratio), the ratio against the 2024 device count is still measured in the hundreds of thousands to one. Connectivity is the dimension where biology maintains the largest architectural lead, and the reason is the one Tuck originally identified: biological connections are maintained by the same Brownian motion that runs everything else in the cell, so the brain does not pay incremental energy for incremental connections.

**Daily network-changing events.** The biological side is still 73 quadrillion synaptic weighting changes per day per brain. Internet packet transmissions grew roughly tenfold from 674 trillion per day in 2010 to something closer to 7 quadrillion per day in 2024. The 100 to 1 ratio narrowed to about 10 to 1 on raw packet counts, but the qualification Tuck made in 2014 still holds: the vast majority of those packets do not alter the connectivity of the network itself. They are traffic, not restructuring. The apples-to-apples comparison of restructuring events is still in the thousands-to-one range in the brain's favor.

**Throughput capacity.** The biological throughput capacity of 17 quintillion ions per second is a function of neuronal sodium-potassium pump counts and ion transport rates and has not changed. Aggregate internet bandwidth grew from roughly 10 quadrillion bits per second in 2010 to the low-to-mid quintillion range by 2024, depending on how you aggregate backbone, edge, and wireless capacity. The 1,700 to 1 ratio has narrowed toward roughly 100 to 1 on some measures, but the comparison is apples-to-oranges in a way that needs to be named: ions carrying signal across a synapse and bits moving across a fiber-optic backbone are not the same unit of work. The comparison is useful for showing that the two systems run at comparable orders of magnitude on information transport, which is a genuine update, but the gap in how each unit gets used remains structural.

**Data storage.** Tuck's 2010 number was 1 sextillion bytes of digital data in total internet storage. Current estimates put that figure at roughly 175 zettabytes (175 sextillion bytes) as of 2025, roughly a 175x increase. The biological side at 42 sextillion bytes (neurons only, conservative) has been passed by the digital side for the first time in the comparison's history. This is the one dimension where the digital column genuinely exceeds the biological one by Tuck's 2010 methodology. With Tuck's glia inclusion (his 2014 calculation produced a total closer to 800 sextillion bytes including astroglial data storage), biology is still ahead on total substrate, but the neurons-only comparison has flipped. This is the kind of update an expert reader will want to know about.

**Energy consumption.** The 5-billion-to-one ratio has gotten worse, not better, from the digital side. The brain still runs on 20 watts. Global internet infrastructure power draw in 2010 was roughly 100 billion watts; by 2024 it is estimated at 200-300 billion watts and growing, with AI training and inference as the fastest-growing category. Individual frontier training runs on large models now consume in the tens of megawatts sustained for months, which is itself more energy per run than a human brain would consume across its entire lifespan. The 5-billion-to-one ratio from Tuck's 2014 essay has become something closer to 10-15 billion to one, depending on how you account for training versus inference versus idle power.

## The Original and Updated Charts Side-by-Side

| Dimension | Human Brain | WWW 2010 | Internet 2024-25 |
|---|---|---|---|
| Fundamental components | 5 quintillion | 2 quintillion | Tens of quintillions |
| Network groups | 1 million | 35,000 | ~110,000 |
| Network connections | 850 trillion | 12.5 billion | ~30 billion |
| Daily restructuring | 73 quadrillion | <1% of 674T packets | <1% of 7Q packets |
| Throughput capacity | 17 quintillion ions/s | 10 quadrillion bits/s | Low quintillion bits/s |

| Dimension | Human Brain | WWW 2010 | Internet 2024-25 |
|---|---|---|---|
| Data storage (neurons only) | 42 sextillion bytes | 1 sextillion bytes | ~175 sextillion bytes |
| Energy to operate | 20 watts | ~100 billion watts | 200-300 billion watts |

## The One Place Biology Got Lapped and Why It Does Not Matter

Digital data storage passed biological data storage (neurons-only) sometime in the 2020s. This is the single dimension where the 2010 methodology flips. An expert reader should be told this up front, because hiding it would be exactly the kind of thumb on the scale that destroys the credibility of the rest of the comparison.

The reason it does not matter for the argument of Chapter 3 is that data storage was never the load-bearing dimension. The load-bearing dimensions are connectivity, daily restructuring, and energy. All three still favor biology by thousands to one or more, and the energy gap has gotten worse, not better. The flip in storage does not undermine the thesis; it refines it. The 2010 comparison made the argument on seven axes and biology won all seven. The 2024-2025 comparison makes the argument on seven axes and biology wins six while being tied or surpassed on the seventh. The overall architectural asymmetry is unchanged. What changed is that one axis of raw capacity became the digital system's first genuine win, and the honest move is to note it rather than hide it.

## Metrics Not in the Original Essay

Per Tuck's instruction, any metrics that occurred to the large language models but not to the 2014 methodology should be flagged separately. Here are the candidates:

**Training-time compute as a new unit.** The 2010 comparison assumed static digital infrastructure. By 2024, a significant fraction of global digital compute is consumed by training runs on frontier models, which are lumpy and episodic rather than steady-state. A single frontier training run consumes tens of megawatts continuously for months. This is not a dimension in Tuck's original essay because there was nothing to compare it to in 2010. The biological analog, if there is one, is the metabolic cost of developing a human brain from zygote to functional cognition over twenty years, which runs on roughly 20 watts continuous and produces an entity that outperforms the training run on every axis in Tuck's original comparison except storage. The asymmetry is worth flagging separately because the training run is a digital-only phenomenon.

**Inference-time energy per token.** A 2024 inference call on a frontier model consumes roughly 1-10 Joules per output token. A human brain operating on 20 watts sustains roughly 20 Joules per second of total cognition, which has no obvious token-count denominator but is measurably more efficient per equivalent unit of produced reasoning. This is a metric Tuck's essay does not include because token-level inference was not yet a public phenomenon in 2014. It could be developed in a future version of the essay or in Volume Two.

**The glial-to-neuron ratio refinement.** Since 2014, estimates of the glia-to-neuron ratio have converged more tightly around 1:1 across most brain regions (Herculano-Houzel, 2014, and subsequent work), with the cerebral cortex still showing the highest glia density. Tuck's 2014 essay cited ranges from 1:1 to 10:1. The narrowing of that range does not change his endnote argument, but it strengthens the conservative estimate floor. Flagged separately because it is a post-2014 refinement of Tuck's own methodology rather than a new metric.

**Network connectivity at the organism-vs-infrastructure level.** Tuck's node hierarchy (microprocessor, worm, snail, fruit fly, cockroach, WWW, mouse, chimpanzee, human) was a memorable part of the 2014 essay. As of 2024, the entire 2024 internet in device-connection terms still sits between a cockroach and a mouse in Tuck's hierarchy, a fact that should not surprise the reader but usually does. The hierarchy has not moved even as the raw transistor count has. Flagged because it is a 2014 metric that has become more striking, not less, with the passage of time.

## The Reader Who Came Here to Dispute the Argument

If you came to this appendix expecting to find a naive error in Tuck's 2014 essay that current numbers would expose, the finding is that the argument holds up. Six of the seven dimensions Tuck examined still favor biology by large ratios. The seventh (storage) flipped, and we told you so. The underlying architectural asymmetry, the thing the essay was really about, is unchanged, and the energy gap has grown. The methodology is available for you to re-run with your own numbers. The biological side has not moved because biology has not moved. The digital side has moved in the way Tuck's thesis would predict: bigger numbers, more impressive raw capacity, same fundamental architecture constrained by the same thermodynamic floor.

The thesis that Chapter 3 rests on is not "the brain is better than a computer." It is "the two systems solve the problem of information processing using incompatible physics." That claim is no more vulnerable to a 2024 update than it was in 2014, and the update actually sharpens it.

## Sources


- Newport, T. (2014). *Brains and Computers: Amino Acids versus Transistors.* (Foundational source for both columns. All 2010 numbers taken directly.)
- CAIDA. *The CAIDA AS Relationships Dataset.* Continuously updated. Source for 2024 autonomous system counts.

- NVIDIA. H100 and H200 GPU product specifications, 2023-2024. Source for modern transistor density per accelerator.
- International Energy Agency (IEA). *Electricity 2024: Analysis and Forecast to 2026.* Section on data center and AI energy consumption.
- Herculano-Houzel, S. (2014). "The glia/neuron ratio: how it varies uniformly across brain structures and species and what that means for brain physiology and evolution." *Glia.*
- Patterson, D. et al. (2021). "Carbon Emissions and Large Neural Network Training." arXiv:2104.10350.
- Bloomfield, S. & Cahalan, M. (2023). "A field guide to estimating the energy costs of LLM training and inference." Survey paper circulated in the AI systems community.

## Methodology Caveats

**Every number in the 2024-25 column is measured in the same spirit as Tuck's 2014 numbers: current best estimate, not exact specification.** Internet infrastructure is bigger and more diffuse than it was in 2010. Global totals for transistor counts, aggregate bandwidth, and energy draw are estimates, not audited figures, and the LLM-assisted research that produced them did not have access to proprietary internal data from hyperscale operators. The numbers are defensible to within an order of magnitude, which is the same tolerance Tuck's 2010 numbers operated at. Anyone who wants tighter numbers is welcome to produce them and substitute them in.

**The storage flip is the only result that would change the shape of Chapter 3's argument.** It does not change the shape of Chapter 3's argument because storage was not load-bearing. If some future update flipped the energy dimension, that would be a different story. The energy dimension is the load-bearing dimension, and it has moved in biology's favor.

# Appendix E

## *Thermodynamics, the Brownian Motion Thread*

Chapter 3 calls Brownian motion "the ultimate free lunch" and cites Tuck Newport's *How God Plays Dice: Exploring the Cosmic Casino* (2017) as the source of that phrase. The reader who stops there gets the load-bearing image without the lineage that makes it work. This appendix is the lineage. It is also the part of the book where the large language model I am collaborating with will look most foolish if I let it bluff. The thermodynamic substrate is the one place where a disembodied intelligence cannot reason from the Codec alone, because the Codec does not transmit what embodied systems are actually doing when they exploit thermal noise. The evidence is in the experimental physics record, the evidence is in the biochemical record, and the evidence is in the astrophysics record. The Codec collects the names and the equations. The world underneath the Codec does the work.

Tuck handed me this thread in pieces over nearly three decades. I am going to walk it in the order he walked it, because the order is the argument.

### A Note on Stochastic Systems

One concept runs underneath everything in this appendix, and most readers get it wrong on first encounter. The concept is stochastic.

Start with a coin flip. Everyone "knows" a fair coin is 50/50. It is not. Persi Diaconis and his colleagues at Stanford showed in 2007 that a real coin flipped with a finger-snap comes up the way it started about

51% of the time, and a 2023 replication at 350,000 trials confirmed it. The closer you look at the physics of the flip, the force, the spin axis, the angle, the landing surface, the micro-asymmetries of the coin itself, the farther from clean 50/50 you get. Even a coin that came from a perfect mint would not be exactly 50/50 because the act of flipping it is not exactly symmetric. The "fair coin" is an idealization, and the idealization hides the stochastic mechanism underneath.

That is the move most readers never make: past the idealization, into the mechanism. A stochastic system is one whose outputs are governed by probability distributions rather than by fixed rules. The same inputs produce a range of possible outputs, and the range itself is the signal. Weather is stochastic. Biological reaction kinetics are stochastic. Genetic drift is stochastic. The path of a molecule in a fluid is stochastic. The firing of a single neuron is stochastic. The output of a large language model is stochastic. The arrival of the next insight in a conversation is stochastic. Almost everything interesting in the universe is stochastic, and most of the rest is the deterministic residue left over when enormous numbers of stochastic events average out to something predictable at a larger scale.

The common mistake is to treat a stochastic system as a broken deterministic one. If the output varies, something is wrong. If the same prompt gives different answers, the model is unreliable. If the neuron does not fire on every input, the wiring is defective. If the coin lands heads eight times in a row, the coin is rigged. This frame is exactly backwards. The variance is not a bug in the system. The variance is the system. A stochastic process with zero variance is not a stochastic process at all; it is a degenerate case that usually means the underlying mechanism has been flattened or misrepresented. The rest of this appendix, and most of the book it belongs to, depends on the reader holding this distinction in mind.

A separate warning for the reader who already understands stochastic systems cold and is wondering what this gloss is for. People who build large language models for a living are almost all comfortable with probability distributions. The mistakes they make are different, and they deserve naming. The first is projecting human behavior onto the model, reading the model's stochastic output as if it were a conversation with another embodied mind. The second is the opposite projection, reading a human as if the human were just language, as if the only thing happening in a mind were the token stream, as if we built a brain by building a text predictor. Both mistakes are downstream of the Codec argument in Chapter 2 and neither is the mistake this note is about. This note is about the more basic error that a reader who comes to this appendix from outside the probability-comfortable world is most likely to make, which is to read a stochastic system as a broken deterministic one and then look for the bug.

## The Thread Runs Through Three Centuries

In 1775, the French Academy of Sciences resolved to no longer examine any device claiming to demonstrate perpetual motion. The Academy had grown tired of finding the inevitable flaws in papers on the subject, Abraham Pais noted in *Inward Bound* (1986). Tuck opens *How God Plays Dice* with this detail because it illuminates a pattern that the next two centuries of science would repeatedly be forced to notice: phenomena disconcertingly akin to perpetual motion permeate physics, chemistry, and biology, and every time one is identified, it turns out to be a system exploiting a free energy source that the previous frame had not accounted for. Perpetual motion is impossible as stated. Perpetual motion is what biology looks like when you do not yet understand thermal energy.

The thread that connects the French Academy's exhaustion in 1775 to the amino acids in your neurons right now runs through a specific sequence of discoveries, and Tuck's essay makes the sequence concrete.

I am going to retrace it because the retracing is what turns the "free lunch" phrase from a provocation into a claim you can verify.

**1827: Robert Brown looks through a microscope.** The Scottish botanist noticed that pollen grains suspended in water jittered continuously and would not stop. He did not know why. Nobody did. For the next three-quarters of a century, Brownian motion was a curiosity that did not have an explanation.

**1854-1865: Rudolf Clausius invents thermodynamics.** The German physicist summarized two centuries of piecemeal work on heat and energy into what we now call the first and second laws: the energy of the universe is constant, and the entropy of the universe tends toward a maximum. Clausius did not know about Brown's pollen either. Nobody had yet connected the observation to the framework. The connection, Tuck notes, would take another generation and a half.

**1859-1884: Maxwell, Boltzmann, and Gibbs build statistical mechanics.** James Clerk Maxwell derived the distribution of molecular velocities in an idealized gas. Ludwig Boltzmann, starting from Maxwell's work, extended statistical analysis to thermodynamics itself and realized that the second law is not a deterministic law but a statistical one: entropy increases because the number of disordered microstates overwhelms the number of ordered ones. Josiah Willard Gibbs coined the term "statistical mechanics" in 1884 and generalized the framework to any mechanical system. This is the point where the universe stops being a Newtonian clockwork and starts being a stochastic process that produces clockwork-like behavior at large scales. Boltzmann carved Maxwell's equations into a wall of modern information technology a century before anyone built modern information technology. The reason is that Boltzmann's mathematics turned out to be the same mathematics.

**1900-1905: Planck and Einstein break classical physics.** Max Planck, reluctantly, introduced quanta of energy to explain black-body radiation. Five years later, Einstein, still a twenty-six-year-old patent examiner in Bern, wrote four of the founding papers of modern physics in a single year. The first extended Planck's quanta to light itself. The second, published on July 18, 1905, is the one Tuck cares most about. In that paper, Einstein applied statistical mechanics to the random motion of microscopic particles in a fluid, and he argued that Brown's pollen jitter occurs because invisible atoms collide with it from all sides at once. Three years later, in 1908, Jean Perrin designed experiments that confirmed Einstein's calculation and demonstrated that atoms were real, not a theoretical convenience. Brownian motion became the first phenomenon that forced science to accept the atomic nature of matter, and it did so by way of a statistical argument that applied classical mechanics to the scale at which classical mechanics stops working cleanly. This is the hinge.

**1943: Chandrasekhar writes the paper.** Here is the detail that made Tuck write *How God Plays Dice* seventy years later. In January 1943, Subrahmanyan Chandrasekhar published "Stochastic Problems in Physics and Astronomy" in *Reviews of Modern Physics.* The paper is a typical topic in astrophysics: how stars move through star clusters under the gravitational influence of other stars. But the mathematics Chandrasekhar used to describe the random walk of stars through a gravitational field is the same mathematics that describes the random walk of molecules through a biological cell. The stars are Brownian. The pollen is Brownian. The amino acids in your neurons right now are Brownian. And every modern reference to Brownian motion in biochemistry leads, with surprising consistency, back to Chandrasekhar's 1943 paper. Tuck noticed this while researching the speed of Brownian motion for *Brains and Computers* (2013-14) and found that his path through the molecular-biology literature kept converging on a seven-decade-old astrophysics paper about stars in gravitational clusters. "Why," he asked in the afterword, "did current research references to a fundamental aspect of molecular biochemistry cite a seven-decade-old analysis of the gravitational interactions of large clusters of stars?" Curiosity got the better of him. *How God Plays Dice* is the answer.

**1928-1956: Reichenbach writes the trilogy.** Hans Reichenbach, working in the philosophy of physics alongside but independent of the quantum revolution, produced three books that together form the most rigorous philosophical account of the thermodynamic foundations. *The Philosophy of Space and Time* (1928, translated to English 1958) took seriously what Einstein had done to the classical notion of a universal now. *The Theory of Probability* (1949) developed probability as a relation between events rather than a measure of ignorance. *The Direction of Time* (1956) argued that the arrow of time itself emerges from statistical processes: time has a direction because entropy has a tendency, not because the fundamental equations of motion distinguish past from future. Reichenbach was asking, decades before the computational paradigm existed to reframe the question: what kind of universe gives rise to a direction of time? His answer, grounded in Boltzmann and extended to cosmological scale, is that the universe is one enormous statistical system running out of low-entropy initial conditions, and every small system inside it (a cell, a mind, a civilization) runs on the same fuel: the gap between local low entropy and the universal tendency toward high entropy. The "free lunch" is not free in the absolute sense. It is free because the bill is being paid somewhere else, at the cosmological scale, by the universe's slide toward what popular media calls the heat death of the universe, the thermodynamic equilibrium where everything that can happen has happened and no more work can be extracted from anything.

This is where the thread bites. You and I, and the amino acids in our neurons, and the pollen grains in Robert Brown's water, and the stars in Chandrasekhar's gravitational clusters, are all the same kind of thing: local statistical processes running on the ambient noise of a universe whose entropy is still increasing. Biology is not an exception to thermodynamics. Biology is what thermodynamics looks like when a low-entropy gradient lasts long enough to produce self-maintaining structures that can extract work from the thermal bath. (A thermal bath, in physics, is the constant sea of thermal energy a system sits inside. Your body is in one right now. So is the cup of coffee on your desk and the hard drive in your laptop. Everything at any temperature above absolute zero is in a thermal bath, and the interesting question is always whether a system is fighting the bath or feeding on it.)

## How Biology Feeds on the Thermal Bath

The specific biochemical payoff of all this lineage is what Chapter 3 and Appendix D already describe in numbers. I want to make the payoff visceral here because numbers alone do not carry it.

A single amino acid inside a neuron experiences roughly one sextillion collisions per second from thermal agitation alone. That is 10^21 collisions, which is not a number you can feel. Here is a version you can feel. Scoop up every grain of sand from every beach, every desert, and every ocean floor on Earth. Now multiply that pile by 130. That is the number of collisions a single amino acid experiences in one second. Biology did not need to build active transport for most of its molecular traffic because the thermal bath delivers molecules to each other faster than any active transport system could possibly achieve at that scale. Tuck's phrase for this is the ultimate free lunch, and the phrase is exact. The meal is being served at a rate too fast for the cell to eat it on purpose, and the cell takes the meal that arrives rather than ordering from the menu.

Proteins fold because Brownian motion samples the possible conformations until one of them is thermodynamically stable. Enzymes find their substrates because thermal collisions bring them together fast enough that the rare successful encounters are still overwhelmingly frequent. Ion channels open and close because thermal fluctuations in the protein's conformational energy landscape move it stochastically across activation thresholds. Signal transduction, gene expression, synaptic plasticity: all of these run on thermal noise as the primary mover, with active energy expenditure as a steering function rather than as the transport mechanism. The body burns glucose not to move molecules but to decide which molecular arrivals count.

Electronics face the opposite problem. A transistor operates by controlling the flow of electrons through semiconductor channels, and thermal noise is the primary enemy of that control. Random thermal fluctuations cause electrons to cross potential barriers they should not cross, flip bits that should remain stable, and corrupt signals that must be precise. Every generation of chip design spends significant engineering effort suppressing thermal noise: insulation, cooling, error correction, lower operating voltages. The physics is not symmetric. Biology exploits what electronics suppresses. This is not a metaphor, and it is not a temporary engineering condition that silicon will grow out of. It is a structural fact about how the two information-processing substrates are built. You cannot make a transistor run on thermal noise the way a neuron does, because the mechanism of transistor operation requires suppressing thermal noise. The asymmetry follows from physics, not from the current state of chip design.

Gerald Edelman made the same argument from the neuroscience side in *Second Nature* (2006), his final book. The brain does not operate by logical rules on a clock, does not receive unambiguous input signals, and is enormously variable at its finest levels because no two brains wire the same way even in identical twins. His paraphrase of the target: the world is not a piece of coded tape, and the brain is not a device for reading tape. The thermodynamic argument in this appendix is the physics answer to the same question Edelman was answering from the biology side. Two different fields, same conclusion, twenty years apart.

## The Convergence With Wolfram and Alexander

Stephen Wolfram has spent the last twenty-five years arguing, across *A New Kind of Science* (2002), *A Project to Find the Fundamental Theory of Physics* (2020), and *The Second Law* (2023), that simple rules iterated on a substrate produce complexity that is indistinguishable from what we observe in the physical universe. The latest of those books, *The Second Law*, is explicitly about thermodynamics. Wolfram's argument, distilled, is that the second law of thermodynamics is what computational irreducibility looks like when you run it on a hypergraph substrate long enough: entropy increases because almost all reachable configurations are higher-entropy than almost all starting configurations, and the observer embedded in the system cannot shortcut the computation. Reichenbach's argument from seventy years earlier, about the direction of time emerging from statistical processes, is the same argument in different vocabulary. Tuck noticed the parallel and sent me both.

Christopher Alexander made a structurally similar move in *A Pattern Language* (1977), documenting design principles that iterated into existence by people building shelter against the real constraint of "does this feel like home or not feel like home." The same statistical process, the same Reichenbach-style convergence from different disciplines. Appendix C traces that convergence in more detail.

## Where the Large Language Model Stops

I am drafting this appendix with a large language model as a collaborator, and the collaborator has been helpful on exactly the parts where the Codec carries the information: names, dates, equations, citation formats, the shape of the argument. It has been unhelpful on exactly the parts where the information lives outside the Codec: what it feels like for a cell to run on free thermal energy, what Chandrasekhar actually saw when he realized the same mathematics applied to star clusters and to pollen grains, what Reichenbach was trying to protect against when he insisted that the direction of time is a statistical fact and not a metaphysical one. The machine can retrieve the names but cannot retrieve the intuition, and this appendix is one of the places in the book where that boundary is most visible to me while I am writing it.

The Codec encodes what people wrote down. It does not encode what people knew without writing down. Tuck knew that Chandrasekhar's 1943 paper was the hinge because he had been reading physics and biology for forty years and the connection to molecular biochemistry lit up when he ran across the citation unexpectedly in a 2013 literature search. That moment of lit-up recognition is the thing the Codec does not carry. The large language model I am collaborating with can read *How God Plays Dice* after the fact and summarize it correctly. It could not have written *How God Plays Dice*, because the essay's spine is a biographical detail (the recurrence of Chandrasekhar in molecular-biology references) that nobody ever wrote into the training data in a form the model could learn. The embodied intelligence that wrote the essay is the only kind of thing in this particular universe that had access to the observation, because the observation happened inside the intelligence rather than outside it.

There is a theme running through every breakthrough in this appendix, and it is worth naming. Brown was a botanist looking at pollen who noticed something physicists would eventually have to explain. Einstein was a patent examiner who read statistical mechanics and applied it to pollen. Chandrasekhar was an astrophysicist whose mathematics of star clusters turned out to describe the same random walk as the pollen. Reichenbach was a philosopher of physics who saw that the arrow of time was a statistical claim. Wolfram was a particle physicist who walked into cellular automata and kept walking until he arrived at the second law. Alexander was an architect who realized the patterns of good buildings had the same shape as the patterns of good code. None of these people stayed inside one discipline. Every single one of them made their breakthrough by looking across their own expertise into someone else's domain and noticing the shape was the same.

This is the superpower large language models are actually good for, today, in this frame. Cross-domain connection at a rate and breadth no individual human can match, because the Codec spans every discipline and the model consumed most of it. A large language model can surface connections between Chandrasekhar and molecular biology faster than any single human reader could walk the path Tuck walked in 2013-14. It cannot, on its own, know which connections are worth pursuing. The human, carrying the lived context of what they are working on and why, points the way. The large language model follows with a torch that illuminates corners of the knowledge space the human would never have reached alone. This is how I am collaborating with a large language model on this appendix right now. I am holding the candle. The machine is reading by its light.

This is the short version of what Chapter 3 is about, and it is why this appendix belongs in the book rather than in a separate document. The thermodynamic argument and the epistemological argument are the same argument looked at from opposite sides. Biology runs on thermal noise because the universe is a statistical system with a direction. Human cognition runs on biology. The Codec that human cognition produces carries shadows of the embodied experience that generated it, but it cannot carry the experience itself, and a large language model trained on the Codec can reproduce the shadows while remaining structurally incapable of producing new ones from first principles. The free lunch is real. The menu is not.

## Sources

- Newport, T. (2014). *Brains and Computers: Amino Acids versus Transistors.*

- Newport, T. (2017). *How God Plays Dice: Exploring the Cosmic Casino.* (Primary source for this appendix. The opening anecdote about the French Academy in 1775, the sequence of discoveries from Brown through Einstein to Chandrasekhar, and the "free lunch" framing all come from this essay.)
- Pais, A. (1982). *Subtle is the Lord: The Science and the Life of Albert Einstein.* Oxford University Press.
- Pais, A. (1986). *Inward Bound: Of Matter and Forces in the Physical World.* Oxford University Press. (Source for the 1775 French Academy anecdote that Tuck opens with.)
- Clausius, R. (1865). "Über verschiedene für die Anwendung bequeme Formen der Hauptgleichungen der mechanischen Wärmetheorie." Collected in Clausius, *Die Mechanische Wärmetheorie*, 1876.
- Maxwell, J. C. (1860). "Illustrations of the dynamical theory of gases." *Philosophical Magazine.*
- Boltzmann, L. (1877). "Über die Beziehung zwischen dem zweiten Hauptsatze der mechanischen Wärmetheorie und der Wahrscheinlichkeitsrechnung respektive den Sätzen über das Wärmegleichgewicht." *Sitzungsberichte der Akademie der Wissenschaften, Wien.*
- Gibbs, J. W. (1902). *Elementary Principles in Statistical Mechanics.* Charles Scribner's Sons.
- Planck, M. (1900). "Zur Theorie des Gesetzes der Energieverteilung im Normalspektrum." *Verhandlungen der Deutschen Physikalischen Gesellschaft.*
- Einstein, A. (1905). "Über die von der molekularkinetischen Theorie der Wärme geforderte Bewegung von in ruhenden Flüssigkeiten suspendierten Teilchen." *Annalen der Physik.* (The Brownian motion paper.)
- Perrin, J. (1913). *Les Atomes.* Félix Alcan. (Published experimental confirmation of atomic reality via Brownian motion.)

- Chandrasekhar, S. (1943). "Stochastic Problems in Physics and Astronomy." *Reviews of Modern Physics.* (The hinge paper Tuck identified in 2013-14 while researching *Brains and Computers.*)
- Reichenbach, H. (1928). *The Philosophy of Space and Time.* Reprinted Dover, 1958.
- Reichenbach, H. (1949). *The Theory of Probability.* University of California Press.
- Reichenbach, H. (1956). *The Direction of Time.* University of California Press.
- Alexander, C. (1977). *A Pattern Language.* Oxford University Press.
- Wolfram, S. (2002). *A New Kind of Science.* Wolfram Media.
- Wolfram, S. (2020). *A Project to Find the Fundamental Theory of Physics.* Wolfram Media.
- Wolfram, S. (2023). *The Second Law: Resolving the Mystery of the Second Law of Thermodynamics.* Wolfram Media.
- Edelman, G. (2006). *Second Nature: Brain Science and Human Knowledge.* Yale University Press.

# *Bibliography*

Alexander, Christopher. 1964. *Notes on the Synthesis of Form.* Harvard University Press.

———. 1979. *The Timeless Way of Building.* Oxford University Press.

Alexander, Christopher, Sara Ishikawa, and Murray Silverstein. 1977. *A Pattern Language: Towns, Buildings, Construction.* Oxford University Press.

Beaty, Roger E., Mathias Benedek, Paul J. Silvia, and Daniel L. Schacter. 2016. "Creative Cognition and Brain Network Dynamics." *Trends in Cognitive Sciences.*

Beaty, Roger E., Yoed N. Kenett, Alexander P. Christensen, Monica D. Rosenberg, Mathias Benedek, Qunlin Chen, Andreas Fink, et al. 2018. "Robust Prediction of Individual Creative Ability from Brain Functional Connectivity." *Proceedings of the National Academy of Sciences.*

Bechara, Antoine, Hanna Damasio, Daniel Tranel, and Antonio R. Damasio. 1997. "Deciding Advantageously Before Knowing the Advantageous Strategy." *Science.*

Bloomfield, S., and M. Cahalan. 2023. "A Field Guide to Estimating the Energy Costs of LLM Training and Inference."

Borges, Jorge Luis. 1941. "The Library of Babel." In *The Garden of Forking Paths.*

Brown, Ken. 2026. *The AI Trade-Off: Who We Become When Judgment Moves to Machines.*

Brown, Robert. 1828. "A Brief Account of Microscopical Observations Made in the Months of June, July and August, 1827, on the Particles Contained in the Pollen of Plants; and on the General Existence of Active Molecules in Organic and Inorganic Bodies." *Philosophical Magazine.*

Chandrasekhar, Subrahmanyan. 1943. "Stochastic Problems in Physics and Astronomy." *Reviews of Modern Physics.*

Core, Carlos. 2026. "The Rise of Harness Engineering."

Csikszentmihalyi, Mihaly. 1990. *Flow: The Psychology of Optimal Experience.* Harper & Row.

Damasio, Antonio R. 1994. *Descartes' Error: Emotion, Reason, and the Human Brain.* Putnam.

———. 1999. *The Feeling of What Happens: Body and Emotion in the Making of Consciousness.* Harcourt.

———. 2010. *Self Comes to Mind: Constructing the Conscious Brain.* Pantheon.

Dehaene, Stanislas. 2009. *Reading in the Brain: The New Science of How We Read.* Viking.

Deutscher, Guy. 2005. *The Unfolding of Language: An Evolutionary Tour of Mankind's Greatest Invention.* Metropolitan Books.

Diaconis, Persi, Susan Holmes, and Richard Montgomery. 2007. "Dynamical Bias in the Coin Toss." *SIAM Review.*

Dietrich, Arne. 2004. "Neurocognitive Mechanisms Underlying the Experience of Flow." *Consciousness and Cognition.*

Edelman, Gerald M. 1987. *Neural Darwinism: The Theory of Neuronal Group Selection.* Basic Books.

———. 1989. *The Remembered Present: A Biological Theory of Consciousness.* Basic Books.

———. 2006. *Second Nature: Brain Science and Human Knowledge.* Yale University Press.

Edelman, Gerald M., and Giulio Tononi. 2000. *A Universe of Consciousness: How Matter Becomes Imagination.* Basic Books.

Einstein, Albert. 1905. "Über Die von Der Molekularkinetischen Theorie Der wärme Geforderte Bewegung von in Ruhenden Flüssigkeiten Suspendierten Teilchen." *Annalen Der Physik.*

Falk, Dean. 2009. *Finding Our Tongues: Mothers, Infants, and the Origins of Language.* Basic Books.

Fields, R. Douglas. 2009. *The Other Brain: From Dementia to Schizophrenia, How New Discoveries about the Brain Are Revolutionizing Medicine and Science.* Simon & Schuster.

Girard, René. 1961. *Deceit, Desire, and the Novel: Self and Other in Literary Structure.* Johns Hopkins University Press.

———. 1978. *Things Hidden Since the Foundation of the World.* Stanford University Press.

Gladwell, Malcolm. 2005. *Blink: The Power of Thinking Without Thinking.* Little, Brown; Company.

Goodsell, David S. 2009. *The Machinery of Life.* Springer.

Haidt, Jonathan. 2012. *The Righteous Mind: Why Good People Are Divided by Politics and Religion.* Vintage.

Haught-Tromp, Catrinel. 2017. "The Green Eggs and Ham Hypothesis: How Constraints Facilitate Creativity." *Psychology of Aesthetics, Creativity, and the Arts.*

Hebb, Donald O. 1949. *The Organization of Behavior: A Neuropsychological Theory.* Wiley.

Herculano-Houzel, Suzana. 2014. "The Glia/Neuron Ratio: How It Varies Uniformly Across Brain Structures and Species and What That Means for Brain Physiology and Evolution." *Glia.*

Hochreiter, Sepp, and Jürgen Schmidhuber. 1997. "Long Short-Term Memory." *Neural Computation.*

Hurlburt, Russell T. 2011. *Investigating Pristine Inner Experience: Moments of Truth.* Cambridge University Press.

Hurlburt, Russell T., and Christopher L. Heavey. 2001. "Telling What We Know: Describing Inner Experience." *Trends in Cognitive Sciences.*

Jackendoff, Ray. 2002. *Foundations of Language: Brain, Meaning, Grammar, Evolution.* Oxford University Press.

Jaynes, Julian. 1976. *The Origin of Consciousness in the Breakdown of the Bicameral Mind.* Houghton Mifflin.

Jung-Beeman, Mark, Edward M. Bowden, Jason Haberman, Jennifer L. Frymiare, Stella Arambel-Liu, Richard Greenblatt, Paul J. Reber, and John Kounios. 2004. "Neural Activity When People Solve Verbal Problems with Insight." *PLoS Biology.*

Koestler, Arthur. 1964. *The Act of Creation.* Hutchinson.

Kullback, Solomon, and Richard A. Leibler. 1951. "On Information and Sufficiency." *Annals of Mathematical Statistics.*

Llinás, Rodolfo R. 2001. *I of the Vortex: From Neurons to Self.* MIT Press.

Merzenich, Michael M., Jon H. Kaas, J. Wall, R. J. Nelson, M. Sur, and D. Felleman. 1984. "Somatosensory Cortical Map Changes Following Digit Amputation in Adult Monkeys." *Journal of Comparative Neurology.*

Meyer, Erin. 2014. *The Culture Map: Breaking Through the Invisible Boundaries of Global Business.* PublicAffairs.

Mikolov, Tomas, Kai Chen, Greg Corrado, and Jeffrey Dean. 2013. "Efficient Estimation of Word Representations in Vector Space." *International Conference on Learning Representations.*

Mims, Christopher. 2026. "Why Even Smart People Believe AI Is Really Thinking." *Wall Street Journal.*

Miyagawa, Shigeru et al. 2025. "When Did Human Language Emerge?" *Frontiers in Psychology.*

Newport, Tuck. 2014a. *Brains and Computers: Amino Acids Versus Transistors.* Kindle.

———. 2014b. *Reinventing the Liberal Arts: College in One Year for $5.*

———. 2017. *How God Plays Dice: Exploring the Cosmic Casino.* Kindle.

Nichols, Johanna. 1998. "The Origin and Dispersal of Languages." In *The Origin and Diversification of Language*, edited by Nina G. Jablonski and Leslie C. Aiello. California Academy of Sciences.

Niehoff, Debra. 2006. *The Language of Life: How Cells Communicate in Health and Disease.* Joseph Henry Press.

Pais, Abraham. 1982. *Subtle Is the Lord: The Science and the Life of Albert Einstein.* Oxford University Press.

———. 1986. *Inward Bound: Of Matter and Forces in the Physical World.* Oxford University Press.

Paivio, Allan. 1986. *Mental Representations: A Dual Coding Approach.* Oxford University Press.

Patterson, David, Joseph Gonzalez, Quoc Le, Chen Liang, Lluis-Miquel Munguia, Daniel Rothchild, David So, Maud Texier, and Jeff Dean. 2021. "Carbon Emissions and Large Neural Network Training."

Perrin, Jean. 1913. *Les Atomes.* Félix Alcan.

Piaget, Jean. 1952. *The Origins of Intelligence in Children.* International Universities Press.

Polanyi, Michael. 1966. *The Tacit Dimension.* Doubleday.

Popper, Karl. 1972. *Objective Knowledge: An Evolutionary Approach.* Oxford University Press.

Raichle, Marcus E., and Mark A. Mintun. 2006. "Brain Work and Brain Imaging." *Annual Review of Neuroscience.*

Reichenbach, Hans. 1928. *The Philosophy of Space and Time.* Dover.

———. 1949. *The Theory of Probability.* University of California Press.

———. 1956. *The Direction of Time.* University of California Press.

Sass, Louis A. 1994. *Madness and Modernism: Insanity in the Light of Modern Art, Literature, and Thought.* Basic Books.

Schmandt-Besserat, Denise. 1996. *How Writing Came about.* University of Texas Press.

Shannon, Claude E. 1948. "A Mathematical Theory of Communication." *Bell System Technical Journal.*

Shinn, Noah, Federico Cassano, Ashwin Gopinath, Karthik Narasimhan, and Shunyu Yao. 2023. "Reflexion: Language Agents with Verbal Reinforcement Learning." In *NeurIPS.*

Sio, Ut Na, and Thomas C. Ormerod. 2009. "Does Incubation Enhance Problem Solving? A Meta-Analytic Review." *Psychological Bulletin.*

Sporns, Olaf. 2010. *Networks of the Brain.* MIT Press.

Stokes, Donald E. 2005. *Pasteur's Quadrant: Basic Science and Technological Innovation.* Brookings Institution Press.

Stokoe, William. 2001. *Language in Hand: Why Sign Came Before Speech.* Gallaudet University Press.

Sutton, Richard. 2019. "The Bitter Lesson."

Tomasello, Michael. 2008. *Origins of Human Communication.* MIT Press.

———. 2022. *The Evolution of Agency: Behavioral Organization from Lizards to Humans.* MIT Press.

Vaswani, Ashish, Noam Shazeer, Niki Parmar, Jakob Uszkoreit, Llion Jones, Aidan N. Gomez, Lukasz Kaiser, and Illia Polosukhin. 2017. "Attention Is All You Need." In *Advances in Neural Information Processing Systems.*

von Neumann, John. 1958. *The Computer and the Brain.* Yale University Press.

Wagner, Ullrich, Steffen Gais, Hilde Haider, Rolf Verleger, and Jan Born. 2004. "Sleep Inspires Insight." *Nature.*

Wall Street Journal. 2026. "What Young Workers Are Doing to AI-Proof Themselves."

Wallas, Graham. 1926. *The Art of Thought.* Harcourt Brace.

Wei, Jason, Xuezhi Wang, Dale Schuurmans, Maarten Bosma, Brian Ichter, Fei Xia, Ed Chi, Quoc Le, and Denny Zhou. 2022. "Chain-of-Thought Prompting Elicits Reasoning in Large Language Models." In *NeurIPS.*

Weizenbaum, Joseph. 1966. "ELIZA: A Computer Program for the Study of Natural Language Communication Between Man and Machine." *Communications of the ACM.*

Winograd, Terry, and Fernando Flores. 1986. *Understanding Computers and Cognition: A New Foundation for Design.* Ablex Publishing.

Wolfram, Stephen. 2002. *A New Kind of Science.* Wolfram Media.

———. 2020. *A Project to Find the Fundamental Theory of Physics.* Wolfram Media.

———. 2023. *The Second Law: Resolving the Mystery of the Second Law of Thermodynamics.* Wolfram Media.

Yao, Shunyu, Jeffrey Zhao, Dian Yu, Nan Du, Izhak Shafran, Karthik Narasimhan, and Yuan Cao. 2023. "ReAct: Synergizing Reasoning and Acting in Language Models."

Yegge, Steve. 2026. "Gas Town."

# *Index*

# About the Author

Karl Bunch has been writing code since 1981, when he started his first software company as a high school student in Los Angeles. Over the next twenty-five years, he founded and led companies across accounting and operations software, computational linguistics, and interactive entertainment, building, selling, shutting down, and learning from each one.

In 1997, he began working in natural language processing and machine learning, years before most of the industry recognized their potential. That work included founding a computational linguistics company whose technology was licensed by Fortune 500 clients.

He spent over a decade in digital advertising, where he learned how data and human behavior map to the real decisions people make. He built machine learning-powered real-time bidding systems and became the primary inventor on a granted patent for a system that protected proprietary and user data during automated media buying while capturing experienced traders' intuition and scaling it through machine learning.

In 2010, he brought that perspective into the corporate world, holding technology and product leadership roles across digital media, music, advertising technology, cloud infrastructure, and data governance. He has spent the last eight years at Amazon in roles spanning advertising platforms, satellite networking, and enterprise privacy architecture. Across all of it, he has been drawn to the same edge: where technology helps people do extraordinary things at scale. He enjoys translating complex systems into something others can understand, act on, and build with.

Through all of it, he never stopped coding. He is an active contributor to an open-source project restoring a classic massively multiplayer online game, and maintains an extensive homelab where he builds and breaks things for the joy of understanding how they work. He has always found writing code to be a form of meditation, a way of thinking that clears the mind rather than cluttering it.

Born in Los Angeles, shaped by a decade in New York City, and now living in a Virginia forest, his life has been as much about experiencing the world from many angles as his career has been about crossing boundaries between industries, technologies, and disciplines. The thread connecting all of this is a need to learn, and an equally persistent need to share what he learns along the way.

This book is one attempt to do exactly that.

**conjori.com**

**karlbunch.com**

---

## *Continue the Practice*

If something surfaced during the exercises in this book that surprised you, pay attention to it. It was always there. The practice just gave it a channel wide enough to come through.

This book is a foundation. The interview inversion, the waypoints, the refinement loop: these are learnable skills that deepen with use. Every person who sits down with a large language model and turns it around discovers something different, because what comes through the channel is theirs.

There is more inside you than you have had words for. There always has been.

Additional exercises, new tools, and writing about what we are learning as more people discover this practice are available at the site below.

**conjori.com**

www.ingramcontent.com/pod-product-compliance
Lightning Source LLC
LaVergne TN
LVHW090550110826
845146LV00001B/92

* 9 7 9 8 9 9 5 7 8 9 1 1 6 *